Edited by Lynda Birke
and Jonathan Silvertown

More than the parts

Biology and politics

First published in 1984 by Pluto Press Limited,
The Works, 105a Torriano Avenue, London NW5 2RX
and Pluto Press Australia Limited, PO Box 199, Leichhardt,
New South Wales 2040, Australia

Cover designed by James Beveridge

Computerset by Promenade Graphics Limited, Cheltenham, Glos

Printed in Great Britain by Photobooks (Bristol) Limited

Bound by W. H. Ware & Sons Limited, Tweed Road, Clevedon, Avon

British Library Cataloguing in Publication Data
More than the parts.
1. Biology—Social aspects 2. Biology
—Political aspects
I. Birke, Lynda II. Silvertown, Jonathan
306'.45 QH333

ISBN 0–86104–607–2

Contents

Steven Rose researches and teaches in the area of brain mechanisms and learning at the Open University. His most recent book, with Dick Lewontin and Leo Kamin, is *Not in Our Genes* (Penguin, 1984). He also edited *Against Biological Determinism* and *Towards a Liberatory Biology* by the Dialectics of Biology Group (Allison & Busby, 1982).
Seán Murphy teaches biology at The Open University and his major research interest is in brain plasticity and the properties of non–neuronal cells in the brain. His recently published book, written with Alastair Hay and Steven Rose, is *No Fire, No Thunder: The Threat of Chemical and Biological Weapons* (Pluto Press, 1984).
Lynda Birke is a feminist and ethologist at The Open University. Her teaching is in biology and women's studies, and her research is into the role of hormones in behavioural development. She is a member of BSSRS, and of the Dialectics of Biology Group, and is currently writing a book on feminism and biology.
Lesley Rogers is a feminist and ethologist at Monash University, Melbourne, where she teaches and does research in pharmacology and animal behaviour. She has written articles on a number of themes, including women's biology, and environmental issues. She is a member of the Dialectics of Biology Group.
Len Doyal is a senior lecturer in philosophy at Middlesex Polytechnic and is the co-ordinator for Science, Technology and Society studies. His current research is in the philosophy of the social sciences and into the problem of human need.
Lesley Doyal is a senior lecturer in sociology at the Polytechnic of North London. She is a feminist and is particularly involved with issues around women's health. She has published *The Political Economy of Health* (Pluto Press, 1979) and *Cancer in Britain: The Politics of Prevention* (Pluto Press, 1983).
The **BSSRS** (British Society for Social Responsibility in Science) **Sociobiology Group** consist of Joe Crocker, Dot Griffiths, Charlie Owen, Allison Quick, Tim Shallice and Helena Sheiham. The group is one of several 'specialist' groups of BSSRS and was set up in the mid-1970s to look at the ways in which theories of sociobiology have implications for politics.
Richard Clarke teaches ecology and science studies at London University Extra-Mural department and is a member of several organizations working to make science serve human need rather than corporate greed.
Jonathan Silvertown is a member of BSSRS and teaches ecology and evolution in the biology department at The Open University. He does research in the evolution of life history and in plant ecology, has published several papers on these subjects and a book, *Introduction to Plant Population Ecology* (Longman, 1982).
Uriel Kitron, Brian Schultz, Katherine Yih and **John Vandermeer** are members of the Ann Arbor Chapter of the New World Agriculture Group, and work in the Division of Biological Sciences at the University of Michigan, Ann Arbor, Michigan, USA.

Introduction

This book contains a collection of articles dealing critically and analytically with several areas of the life sciences. 'Life sciences' is used here to encompass plant and animal biology as well as medicine. An obvious question with which to begin is, of course, why the life sciences? Biological ideas are becoming increasingly important, impinging on our lives at many points, and it is vital for us to try to make sense of them: what effects will they have on our lives or the lives of others? Some of the effects of these ideas may be desirable; we may feel others are of dubious benefit. If we are to do something about the less desirable consequences, however, we have first to understand what those consequences might be.

The life sciences have always had an appeal to those anxious to explain the current state of human nature and society. Explanations which assume that human behaviour is somehow innate, or built in, appeal to some because of the apparent simplicity of such ideas and because they limit the possibility of social change. For some – notably those who stand to benefit from present social arrangements – change is something to be avoided, and comfort can be taken in the thought that we are as we are because of our biology.

During the recent period of recession and economic crisis, these biologically-based arguments have gained greater currency. More importantly, ideas that certain features of human behaviour may be innate have been specifically articulated by the New Right. Richard Verrall, for example, writing in the fascist journal *New Nation* suggested: 'It is the modern science of sociobiology that has finally buried Marxism; it only remains to be seen how long the liberal era of utopian delusions will last and how quickly our obsession with creating the equal society will fade away.' Notions that we have a dislike of strangers coded into our genes have been rapidly espoused by those wishing to defend racism and right-wing immigration policies.

Similarly 'biological' arguments have been put forward against other delusions of creating a more equal society. Steven Goldberg, for example, argued that male domination of society was a direct product of male biology: it was man's hormones that made him more aggressive, and led inevitably to patriarchy. Feminist hopes for a better future would, then, be in vain.

Moreover, such arguments also serve to divert attention from the possibility of social change by a process of victim-blaming. If a particular individual behaves in ways that society evaluates as undesirable, then positing biological causes within that individual puts the blame on her or him. That society itself may be at fault is a question thereby evaded. Perhaps one of the nastiest examples of such biological victim-blaming is the notion that women who are repeatedly assaulted and beaten up, are so because their hormones turn them into passive or even willing victims (see chapter 3). The complex social environment that may have contributed to that woman finding herself in a situation in which she is repeatedly assaulted is, it would seem, irrelevant and unchallengeable; far easier to blame the poor victim herself.

An important point about biologically determinist explanations of human behaviour is precisely that they serve to justify and defend existing social arrangements. We would be unwise, it is sometimes said, to try to change these arrangements, because to do so would be to go against our biology. We will thus never be able to transcend class, race, or gender barriers. This pessimistic vision, with its complete disregard of other factors contributing to these barriers, gives cause for criticism of certain areas of the life sciences. Human sociobiology, for example, emphasises the primacy of genetic determinants of human behaviour, with social or political factors as secondary or even themselves caused by the genes. This additive view of nature, which seeks to explain natural phenomena in terms of environmental components *added on to* genetic components is a pervasive one.

This view is, however, misleading, since it is in practice impossible to separate 'genetic' and 'environmental' components. From the moment of conception, an individual organism is the product of a continuing dialectical interaction between its biology and the environment in which it lives and moves. An important feature of viewing development in terms of interaction is that it stresses that the biology is not necessarily primary; and more significantly, it allows for the possibility that the biology itself might be changed in this process of interaction. It is this interactive view of biology which is emphasized here, implying that the whole is more than the sum of its parts (hence the title of this book).

Nonetheless, the additive model persists, both in academic and popular articles. It has been criticised often, and most of the chapters in this book are rooted in that criticism: they start from the view that human (or indeed animal) lives cannot be understood in terms of a primary, unchanging, and *determining* biology. The additive view remains attractive and in part this may be due to its simplicity – it is, after all, easier to grasp the idea of 'A plus B' than it is to think about interactions between A and B. But also the dominance of the additive view is facilitated precisely because it can readily contribute to the prevailing ideology. If the biology is primary and unchanging, then the status quo can be justified in terms of that underlying biological imperative.

This book is concerned with outlining some current issues in the life sciences, as well as dealing with critiques of the various ways in which this ideology is supported within contemporary biology. However, it is not only at the level of theory/ideology that the life sciences might be criticised. Biological knowledge does not simply produce abstract ideas; many of those ideas can also be applied, particularly in medicine or agriculture. The technology that is thereby generated might additionally then be used in ways that reinforce the dominant biology. The idea that a woman who is a victim of violence is a victim of her biology meets criticism in this book not only because of the inherent ideology but also because the accompanying suggestion is that a 'technical fix' is possible. If her biology is at fault, give her drugs to cure it – far easier than attempting to change social conditions. Similarly, ideas that schizophrenia is caused by a biological fault leads to its treatment by a battery of powerful drugs rather than any alternative modes of treatment (see chapters 2 and 4). Such 'technical fixes' not only do nothing to change the social situations that led to the problem in the first place, but also serve to perpetuate the notion that in the end, it is the individual (and her/his biology) that is at fault.

This book, then, sets out to criticise in different ways several areas of the life sciences. Although some of the criticisms can, in fact, be made about any other science, it is particularly relevant to subject the life sciences to critical analysis because biology – more than other major areas of natural science – addresses directly the question of what we are. Although reductionism *in principle* assumes that biological phenomena can ultimately be explained in terms of phenomena at the subatomic level (see chapter 1), we are still far from attempting any such explanation. At present, a more pressing problem is the extent to which social or political phenomena are reduced to and explained in terms of biology.

Faced with the prevalence of biologically based assessments of human nature, what can we do? Clearly, one response is simply to accept them and do nothing. However, if it is recognised that such assessments do little to enhance our understanding and, indeed, enhance the interest of more powerful social groups over those of others, then it is necessary to criticise. This can take a variety of forms. One response, common in the liberal academic tradition, is simply to assume that the life sciences (like the rest of science) still embody the pure pursuit of truth; if society (or part of society) uses that knowledge in particular ways the blame cannot be laid at science's door. This is the use/abuse view of science which sees scientists as somehow separate from the rest of society, producing apparently value-free facts. If society then chooses not to put these facts to 'good' use, but chooses rather to abuse scientific knowledge for its own less beneficial ends, then that is not the responsiblity of the scientists. In the use/abuse view, scientists are absolved of any responsibility for the consequences of their science.

Another, related, response is to criticise various parts of the life sciences on their own terrain, since biologically determinist arguments are frequently based on science that is poor even in its own terms. This kind of criticism is usually couched in terms of the methodology employed, the ways statistics are used, the inferences drawn, and so on. Though important, there is a danger in such an approach, of implying that somehow making 'better' science, and removing the methodological or other biases will *ipso facto* make the problems go away. The problems of biological determinism do not result simply from 'bad' science.

However, it does remain necessary to carry out criticism of science on a number of different levels. The radical critiques of science, which emphasise the nature of science as a social activity, are important sources of criticism. Nevertheless, a central problem with these is that they will not normally be read by a significant proportion of the people who continue to believe in, and to act upon, those very determinist ideas. For this reason, it is necessary to criticise these ideas where they rest on bad science as widely as possible, including in academic journals likely to be read by those who espouse such ideas.

A major concern of this book, is to provide a broad critique of some areas of the life sciences to illustrate the extent to which deterministic and mechanistic ideas permeate modern biology. Within such a broad critique criticism of 'bad science' can be found, but it is not of central importance. Rather, the authors have sought both to criticise the science itself *and* to locate that science within a broader social framework. They reject the use/abuse model as naive, and start from the assumption that life science (like the other sciences) does not simply *lend* itself to the maintenance or use of particular ideologies, but actually embodies them, both in its practice and in its theories. The practice of biology does not, and cannot, take place in a social vacuum. Thus, biologists frequently use the concepts and language of the everyday social world in which they live, to provide a framework of explanation, as several chapters of this book illustrate (but see especially chapter 6 on sociobiology). Since that social world is clearly far from being an equal society, it is scarcely surprising that the examples provided by the life sciences tend to reflect those inequalities. We are not likely to find life sciences that match the 'utopian delusions' of which Richard Verrall wrote so disparagingly, because the science we have does not exist in anything like an equal society.

It is necessary to criticise the assumptions and ideology of existing life sciences – and indeed, this is the primary task of this book. There is, however, a need to move beyond criticism. There are a variety of ways in which this might be done, but two general approaches are worth brief mention here. One is to attempt to formulate a less reductionist approach to biology itself. This task, while urgent, has the difficulty that one is trying to create something more progressive in the absence of appropriate social change – which may by definition be impossible. Nonetheless, the effort may be fruitful if only to clarify what we would like to see replacing the reductionist, mechanistic biology of which we are so critical here. Moving beyond criticism should also take the form of creating alternative practices. Chapter 10 of this book deals particularly with the attempt of one radical group, in collaboration with agricultural workers, to make agricultural research serve the interests of farm labourers in the American Mid-west, and in Nicaragua.

The book begins with a chapter by Steven Rose, who outlines the history of reductionism in biology and argues its poverty as a mode of explanation. This is followed by two chapters which outline specific cases of biological reductionism: in neuroscience (Séan Murphy) and in the study of hormones and behaviour (Lynda Birke). These chapters illustrate the problems and consequences of reductionist explanations, particularly when the reductionism provides a framework for the treatment that patients receive within medicine.

These are followed by two chapters which continue the theme of reductionism in medicine. Lesley Rogers's chapter is concerned in critically analysing some of the factors which contribute to increasing rates of drug prescription, and some of the assumptions which underlie pharmacological research. The next chapter, by Len Doyal and Lesley Doyal looks more broadly at the assumptions and priorities of Western scientific medicine. Having described the failings and problems of the mechanistic medical model which is now dominant, they suggest that what is needed is a more ecological and holistic approach to both disease *and* health. These chapters deal with reductionism at the level of the individual: that is, with the dominant assumption (for example within scientific medicine) that if the individual or some part of her/him misbehaves, then it is the fault of some feature of the individual's biology. In contrast, chapters 6 and 7, both of which are written by the Sociobiology Group of BSSRS (the British Society for Social Responsibility in Science), deal with human sociobiology, and with the concept of dominance in ethology. These subjects are less concerned with the quirks of any deviant individual, and more with the *norms* – or perceived norms – of behaviour, although these norms may then be explained in terms of the selfish action of the individual acting to perpetuate her/his genes.

These are followed by two chapters on ecology. Richard Clarke writes about concepts of population in relation to politics, while Jonathan Silvertown looks at the history of the idea of interspecific competition in the science of ecology. It is quite evident from ecology's history that its central problems have often been set by the ideological climate of the time. This, in turn, has then had material consequences in terms of, for example, population policies. Chapters 10 and 11 have in common a concern with what happens as a consequence of particular kinds of research, although they deal with very disparate areas. Chapter 10 is by Uriel Kitron, Brian Schultz, Katherine Yih, and John Vandermeer, and deals with agricultural research. One of the concerns of this chapter is what happens to people as a result of the organisation of agricultural research and technology; however, a more central concern of this chapter is with the creation of alternatives that serve the interests of those directly concerned with the products of the research – the agricultural workers themselves.

The final chapter in a way comes full circle. Here, the concern is with the implications of the reductionist model in the ways that we treat animals. Although other chapters of this book deal specifically with questions raised about *human* behaviour, the issue of our relationship with other species is increasingly a source of concern. In this chapter, Lynda Birke outlines briefly some of the ambiguity of our relationship with animals and relates it to the mechanistic view of nature that has been dominant since the beginning of the Scientific Revolution. So, where we started this book by considering some of the history of that view, we come finally to return to it in a somewhat different vein. It is that mechanistic, reductionist view of the world that we have criticised, in various ways, throughout the book. We would hope that that criticism provides a base from which to change it, and in the final conclusion we consider some changes that are being made.

We are conscious of two subject areas in the life sciences that are of topical importance but which we have not covered in this book. One is biotechnology – the new buzzword for all biologically-based industries from brewing to forestry. The most rapidly advancing area of biotechnology is, of course, genetic engineering which has been the subject of a recent critique by Ed Yoxen.[1] Our other conscious omission is the resurgence in creationism, particularly in the USA, where it has developed as a part of the anti-intellectual armoury of the right. Though creationism cannot be ignored as a political movement which is distinctly reactionary and anti-science, it is not itself part of biological science. The subject has been amply dealt with by Kitcher.[2]

1. Biological reductionism: its roots and social functions

Steven Rose

Ask biologists what they are trying to do when they record data, make experiments or propose theories. There are obvious answers, like earning a living, solving a problem set by the research director or completing a PhD. Go beyond those and the answer is likely to be that they are trying to find out how or why a particular phenomenon occurs in an ecosystem, a living organism or a subcellular preparation. But what sort of answer is expected to questions of why? and how? What would be regarded as a satisfactory explanation of a particular biological event?

This chapter will consider the answer to this question. It is argued that modern biology seeks a particular *reductionist* type of explanation of phenomena. Suggestions are put forward as to how this reductionist domination arose and the errors which it induces. Finally, there are pointers towards a non-reductionist alternative.[1]

Biology is a strange amalgam of different disciplines. The organism that the biochemist or the physiologist or the geneticist studies may well be the same: the differences between these disciplines lie in the questions they ask of the organism and the level of analysis at which they study it. For the biochemist, the organism is a sort of sack of molecules to be extracted and purified, and their composition and interactions determined. To the physiologist, the organism is a mass of cells organised into discrete but interacting organs; for the ecologist, it is not the individual but the population of organisms that is important. The geneticist asks why one organism is different from another of the same species; the developmental biologist wonders how the organism grows from a single cell or embryo into its characteristic adult form. What is the relationship between all these different questions and levels of analysis?

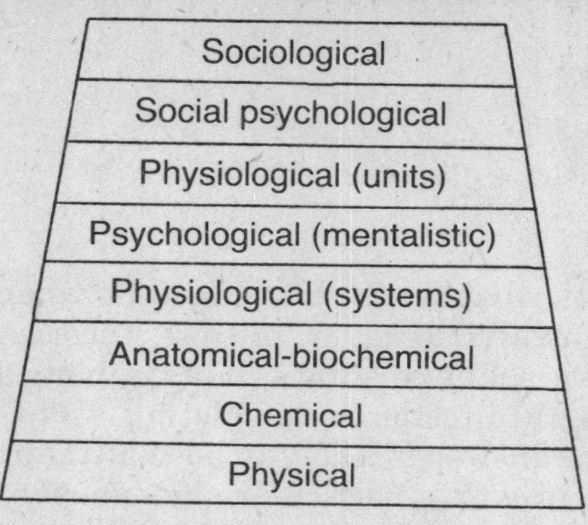

Fig. 1 The conventional hierarchy of science

It is conventional to speak of levels of analysis in biology as if they were arranged in a hierarchical order, as in Fig. 1. Going downwards through the hierarchy there is a movement, so the convention claims, in the direction of increasingly fundamental components, towards the atomic and the subatomic; whereas in moving upwards, the worlds became increasingly complex up to the social domain of the interactions of individuals and populations. Given such a hierarchy, how can we set about explaining a phenomenon at any given level?

Take a simple example: the contraction of a muscle in a frog's leg. There are possibly five types of answers that biologists might give to the question, 'What caused the contraction? Their relationship is summed up in Fig. 2.

'Within-level' explanations

One might, for example, say that the frog's muscle twitched *because* an appropriate set of impulses passed down the motor nerve innervating it, which signalled the instruction to contract. This sort of explanation will describe a present phenomenon as *caused* by an immediately prior event. First the nerve fires, and then the muscle twitches – and we can

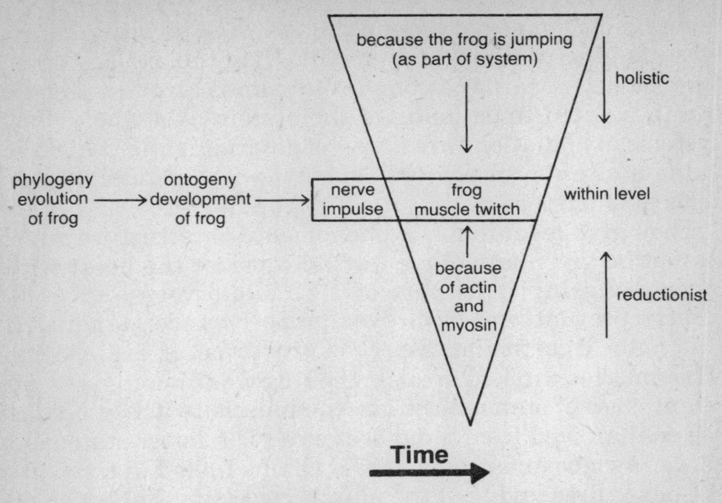

Fig. 2 The relationship between the types of biological explanation

go on to explain that the nerve fired as a result of some earlier appropriate set of inputs to its motor neurons, derived from the frog's brain and/or its sensory input. So we have a sequential series of events that follow one another in time and are linked in a transitive and irreversible way. That is, *first* event A occurs; *then* as a result, event B; and as a result, event C, and so on. This is a straightforward causal chain, with all the individual components described in the same language and within a single level of analysis – that of the physiologist. The sequence can be summarised thus:

external events → sensory stimulation → brain integration → motor nerve firing → muscle twitch

The single headed arrows emphasise that the sequence cannot run backwards, that is, the muscle twitch cannot cause firing of the motor nerve.

'Top–down' explanations

One can, however, consider the activity of the whole organism, and then state 'the muscle twitched *because* the frog was jumping to escape a predator'. Here the explanation of the activity of part of a complex system is given in terms of the integrated functioning of the system as a whole. Such systems explanations are a way of describing the goal directedness of an organism (teleonomy) without implying the concept of purpose.

They give meaning to a phenomenon or structure which cannot be given any other way; to consider the heart without recognising it as part of a circulatory system would clearly prevent one from ever properly understanding it. But note a particular aspect of this type of explanation. 'The muscle twitched *because* the frog was jumping' is a different class of statement from 'the muscle twitched *because* a signal arrived from a motor nerve'. The latter statement, as we saw above, is about causal chains linked in time: *first* the nerve fires and *then* the muscle contracts. But we do not say *first* the frog jumps and *then* the muscle twitches. The two events are not sequential; rather the use of the word 'cause' here implies a logical connection, not a temporal one. Normally frogs are unable to jump unless during this jump, their leg muscles twitch; twitching is part of the activity of jumping. So the types of top–down explanation we can offer are different in kind from those within a single level.

But there are problems in integrating top–down explanations, or *holistic* as they are generally known, with other types of explanation, and there is much confusion about them. Some biologists have regarded them as quite improper, because of the confusion of the two senses in which the word 'cause' is used. Others claim that only holistic explanations can really be satisfactory, that there is a process of 'downward causation' by which the properties of the system – the organism – *constrain* or *determine* the behaviour of the parts. The system becomes thus 'more important' than the parts of which it is composed. If an experimenter severed the motor nerves to the frog's leg muscle, or paralysed the muscle with a chemical poison, the frog would still endeavour to escape its predator, possibly successfully, by employing a different set of muscles or a different escape strategy. To the goal-directed organism there are multiple paths to a given end.

'Bottom–up' explanations

Holism bears a sort of mirror-image relationship to the type of explanation which is the main topic of this chapter, that of *reductionism*. Consider the frog muscle. It is itself composed of individual muscle fibres. These themselves are largely composed of fibrous proteins. In particular, there are two types of protein, actin and myosin, arranged within the muscle fibril in characteristic arrays. When muscle fibrils contract, the actin and myosin chains interdigitate; the conformational change as they slide between one another involves the expenditure of energy, and a chemical substance, ATP, is broken down in the process. So a bottom–up explanation of the muscle twitch would be in terms of the proteins of which the muscle is composed. It contracts *because of* the protein filaments sliding past each other. It is possible to go on to explain the conformational changes in these proteins in terms of the amino acid composition of the individual actin and myosin molecules, and the molecular interactions of group transfer molecules like ATP. These in turn could be described in terms of the atomic or quantum structure of the molecules, and so on.

Like holism, then, reductionism is a between-levels explanation. It is the most commonly accepted type of explanation offered by biologists. The scientific culture of contemporary Western society, with its belief in the hierarchy of sciences of Fig 1, teaches that reductionism offers 'more fundamental' explanations than any other. Higher order levels are, it is argued, in the long run (if not at present) to be explained in terms of the lower order levels.

Indeed, some reductionists go on to claim that *the* task of science is to dissolve the higher orders completely into the lower orders. This type of thinking is particularly strong among biochemists and molecular biologists like Jacques Monod and Francis Crick. Francis Crick has argued that all the important biological questions can be 'solved' by concentrating on unravelling the molecular architecture and dynamics of the bacterium E.coli. It was Crick who formulated what he described as the 'central dogma' of molecular biology, the unilinear chain of causation, which says that information flows only *from* the genetic material, from DNA to protein, but not in the reverse direction.

Reductionism is a powerful explanatory principle in biology, and we are easily led to believe that such reductionist explanations are the most important, or indeed that they are the only 'real' scientific ones. Most of us find the idea that the movement of actin and myosin sliding past one another is the *cause* of the muscle contraction less difficult than the idea that the contraction was *caused* because the frog was attempting to escape from a predator. Many cell biologists would feel uneasy with the latter teleonomic form. Yet formally, reductionist 'causation' is of the same type as holistic 'causation' and both are unlike within-level causation. That is to say, there are not *two* sequential events, *first* the passage past one another of the actin and myosin chains and *then* the muscle contraction. Both happen simultaneously. The passage of the actin and myosin chains past one another is just another way of describing muscle contraction. There is a single event or phenomenon which can be described at one of two different levels. The muscle contraction is *identical* to the sliding of the actin and myosin chains. Thus the relationship between the two descriptions, of muscle contraction and of actin and myosin chains sliding past one another, is not *temporal*, as offered by the within-level explanations; rather it is one of *logical necessity*. The muscle fibril is an ensemble of actin and myosin molecules, and when it contracts the configuration of these molecules alters, by definition.

Developmental and evolutionary explanations

Before looking at the origins and social function of reductionism, it is necessary to refer briefly to the two other types of explanation of the frog muscle twitch shown in Fig. 2. Developmental explanations would attempt to explain the twitch of the muscle in terms of the sequence of events by which muscle cells become specialised from the early embryo and by which they become attached at the two ends and innervated by appropriate nerves, so that the specificity of nerve–muscle connections is achieved and maintained, and the twitch then produces certain defined movements. Such explanations are concerned with how the orchestration of various muscles and nerves is achieved so that their actions are co-ordinated into an appropriate output. A developmental explanation differs from a physiological one in that it tells an historical story; the causal sequence begins not merely with the firing of the motor nerve but before the birth of the organism. Hence the time-scale of the explanation offered is much longer and the purpose it must then serve in terms of accounting for a postulated activity in terms of the unrolling past of the organism is very different.

Evolutionary explanations are concerned also with temporal sequences of events and historical causations for a present situation, but compared with developmental explanations, the time sweep of such attempted explanations transcends the lifetime of any individual and looks at that of the species as a whole. Such explanations have been very popular in recent years, and they too have been subject to certain misconceptions (see chapter 6) especially about the significance of the term 'adaptation' to which we will have need to return in due course.

The origins of reductionism

Having surveyed the five types of explanations offered by contemporary biology, I now return to the question of the origins of reductionist thought. My thesis is that it arose with the birth of modern science in seventeenth century Europe, and that its history is intimately connected with the development of a particular world view. The rise of modern physics, first with Galileo and then particularly with Newton, ordered and atomised the natural world. Beneath the surface world in all its infinite variety of colours, textures, and varied and transient objects, the new science found another world of absolute masses interacting with one another according to invariant laws which were as regular as clockwork. Causal relationships linked falling bodies, the motion of projectiles, the tides, the moon, and the stars. Gods and spirits were abolished or relegated merely to the 'final cause' which set the whole clockwork machinery in motion. (Actually Newton himself remained both religious and mystic throughout his life, but that is one of the minor quirks of personal history: the effect of Newtonian thought was the reverse of Newton's personal philosophy.) By contrast with the feudal world, the universe thus became de-mystified and, in a manner, disenchanted as well.

The post-Newtonian world that emerged was one in which once again heavenly and earthly orders were in seeming harmony. The new physics was dynamic and not static, as were the new processes of trade and exchange which came with the development of the new capitalist economic system. There was a set of new abstractions to describe the world in which a series of abstract forces between atomistic and unchanging masses underlay all transactions between bodies. Drop a pound of lead and a pound of feathers from the leaning tower of Pisa, and the lead will arrive at the ground first because the feathers will be more retarded by air pressure, frictional forces, and so on. But in Galileo's and Newton's equations, the pound of feathers and the pound of lead arrive simultaneously because the *abstract* pound of lead and pound of feathers are equivalent unchanging masses to be inserted into the theoretical equations of the laws of motion.

These abstractions paralleled the world of commodity exchange in which the new capitalism dealt. To each object there are attached properties of mass or value, which are equivalent to or can be exchanged for objects of identical mass or value. Commodity exchange is timeless, unmodified by the frictions of the real world; for example, a coin does not change its value by passing from one hand to another, even if it is slightly damaged or worn in the process. Rather, it is an abstract token of a particular exchange value. It was not until the nineteenth century that this view could become fully dominant. Joule demonstrated that all forms of energy and heat, electromagnetism and chemical reactions were interchangeable and related by a simple constant, the mechanical equivalent of heat; and later Einstein demonstrated the equivalence of matter and energy. These demonstrations corresponded to an economic reductionism whereby all human activities could be assessed in terms of their equivalents in pounds, shillings and pence. Nature and humanity itself had become a source of raw materials to be extracted by applying science and technology – which expertise remained firmly in the hands of the dominant class and gender. The transition from the pre-capitalist world of nature could not be more complete.

So far physics has been discussed as though it were all of science. But where did the new mechanical and clockwork vision of the physicists leave the status of living organisms? Just as modern physics starts with Newton, so modern biology must begin with Descartes, philosopher, mathematician and biological theorist.

For Descartes the world was machine-like, and living organisms merely particular types of clockwork or hydraulic machines. It is this Cartesian machine image which has come to dominate science and to act as the fundamental metaphor legitimating the bourgeois world view of a mechanical nature. That the machine was taken as a model for the living organism – and not the reverse – is of critical importance. Bodies are indissoluble wholes that lose their essential characteristics when they are taken to pieces. Machines, on the contrary, can be disarticulated to be understood and then put back together. Each part serves a separate and analysable function, and the whole operates in a regular, law-like manner that can be described by the operation of its separate parts impinging on each other.

Descartes' machine model was soon extended from non-human to human organisms. It was clear that many, in fact most, human functions were analogous to those of other animals and, therefore, were also reducible to mechanics. However, humans had consciousness, self-consciousness and a mind, which for Descartes, a Catholic, was a soul and by definition the soul, touched by the breath of God, could not be a mere mechanism. So there had to be two sorts of stuff in nature: matter, subject to the mechanical laws of physics, and soul or mind, a non-material stuff which was the consciousness of the human individual, their immortal fragment. How did mind and matter interact? By way of a particular region of the brain, speculated Descartes, the pineal gland, in which the mind/soul resided when incorporate, and from which it could turn the knobs, wind the keys, and activate the pumps of the body mechanism.

So developed the inevitable but fatal dysjunction of Western scientific thought, the dogma known in Descartes' case and that of his successors as 'dualism', but one which, as we shall see, is the inevitable consequence of any sort of reductionist materialism which does not in the end wish to accept that humans are 'nothing but' the motion of their molecules. Dualism was a solution to the paradox of mechanism which would enable religion and reductionist science to stave off for another two centuries their inevitable final contest for ideological supremacy. It was a solution which was compatible with the capitalist order of the day because in weekday affairs it enabled humans to be treated as mere physical mechanisms, objectified and capable of exploitation without contradiction, while on Sundays ideological control could be reinforced by the assertion of the immortality and free will of an unconstrained incorporeal spirit unaffected by the traumas of the workday world to which its body had been subjected.

The development of a materialist biology

For the confident and developing science of the eighteenth and nineteenth centuries, dualism was but a stepping stone towards a more thoroughgoing mechanical materialism. The demonstration by Lavoisier that the processes of respiration and the sources of living energy from the oxidation of foodstuffs in the body tissues, were exactly analogous to those of the burning of a coal fire, was perhaps the most striking vindication of this approach. It was the first time that the programmatic statement that life must be reducible to molecules could be carried into practice.

But progress in the identification of body chemicals was slow. The demonstration that the substances of which living organisms are composed are only 'ordinary' albeit complicated chemicals came early in the nineteenth century.

The intractability of the giant biological molecules – proteins, lipids, nucleic acids – to the analytical tools then available remained a stumbling block. The mechanists could make programmatic statements about the reducibility of life to chemistry, but these remained largely acts of faith. It was not until a century after the first non-organic synthesis of simple body chemicals that the molecular nature and structures of the giant molecules began to be resolved (and really not until the 1950s that progress became very rapid). The last remaining notion that there would be some special 'life-force' operating among them which distinguished them absolutely from lesser, non-living chemicals lingered until the 1920s.

Nonetheless, a radically reductionist programme characterised the statements of many of the leading physiologists and biological chemists of the nineteenth century. In 1845, four rising physiologists, Helmholtz, Ludwig, Du Bois Reymond, and Brucke, swore a mutual oath to account for all bodily processes in physicochemical terms. They were followed by others: for instance, Moleschott and Vogt, thoroughgoing mechanical materialists who claimed that humans are what they eat, genius is a question of phosphorus and that the brain secretes thought as the kidney secretes urine. Virchow, one of the leading figures in the development of cell theory, was also part of a long tradition of social thought which argued that social processes could be described by analogy with the workings of the human body.

It is important to understand the revolutionary intentions of this group. They saw their philosophical commitment to mechanism as a weapon in the struggle against orthodox religion and superstition. Several of them were also militant atheists, social reformers, or even socialists. Science, they believed, would alleviate the misery of the poor and strengthen the power of the state against the capitalists, and even, in some measure, help democratise society. Their claims were part of the great battle between science and religion in the nineteenth century for supremacy as *the* dominant ideology of bourgeois society, a fight whose outcome was inevitable but whose final battlefield was to be that of Darwinian natural selection rather than physiological reductionism.

The best known philosopher of the group was Feuerbach, and it was against his version of mechanical materialism that Marx launched his famous theses.

The theses on Feuerbach proved the starting point for Marx's own, and more explicitly Engels' long-running attempts to transcend mechanical materialism by formulating the principles of a materialist but non-reductionist account of the world and humanity's place within it: dialectial materialism. But within the dominant perspective of biology in the Western tradition, Moleschott's mechanical materialism was to win out, stripped of its millenarial goals and, by the late twentieth century, revealed as an ideology of domination. When biochemists today claim that 'a disordered molecule produces a diseased mind', or psychologists that inner-city violence can be cured by cutting out sections of the brains of ghetto militants, they are speaking in precisely this Moleschottian tradition.

To complete the mechanical materialist world picture, however, a crucial further step was required involving the question of the nature and origin of life itself. The mystery of the relationship of living to non-living presented a paradox to the early mechanists. If living beings were 'merely' chemicals, it should be possible to recreate life from an appropriate physico-chemical mix. Yet one of the biological triumphs of the century was the rigorous demonstration by Pasteur that life only emerged from life; spontaneous generation did not occur. The resolution of this apparent paradox awaited the Darwinian synthesis, which was able to show that although life came from other living organisms and could not arise spontaneously, each generation of living things might change and evolve as a result of the process of natural selection.

With the theory of evolution came a crucial new dimension to the understanding of living processes, the dimension of time. Species were not fixed immemorially but were derived in past history from earlier forms. Trace life back to its evolutionary origins and one could imagine a primordial warm chemical soup in which the crucial chemical reactions could occur. Living forms could coalesce from this pre-biotic mix. Darwin speculated about such origins, although the crucial theoretical advances depended on the biochemist Oparin and the biochemical geneticist Haldane in the 1920s, both, incidentally, consciously attempting to work within a dialectical and non-mechanist framework. Experiments only began to catch up with theory from the 1950s onward.

The consequence of Darwin's theory of evolution was to finally change the form of the legitimating ideology of bourgeois society. No longer able to rely upon the myth of a deity who had made all things bright and beautiful and assigned each to their estate, the rich ruler's castle or the poor peasant's gate, the dominant class de-throned God and replaced him with science. The social order was still to be seen as fixed by forces outside humanity, but now these forces were natural rather than deistic. If anything, this new legitimator of the social order was more formidable than the one it replaced. It has, of course, been with us ever since.

Natural selection theory and physiological reduction were explosive and powerful enough statements of a research programme to occasion the replacement of one ideology, religion, by another – a mechanical, materialist science. They were, however, at best only programmatic, pointing along a route which they could not yet trace. For example, in the absence of a theory of the gene, Darwinism could not explain the maintenance of inherited variation that was essential for its operation. The solution awaited the development of modern genetical theory which took place at the turn of the twentieth century. This in turn produced the neo-Darwinian synthesis of the 1930s and the recurrent attempts to parcel out biological phenomena into discrete and essentially additive causes, genetic and environmental – the science of biometry.

The central dogma: centerpiece of the mechanist programme

Mechanist physiology took even longer to triumph. It depended on the development of powerful new machines and techniques for the determination of the structure of the giant molecules, for observing the microscopic internal structure of cells, and, above all, for studying the dynamic interplay of individual molecules within the cell. By the 1950s it had begun to be possible to describe and account for (in the mechanistic sense) the behaviour of individual body organs – muscles, liver, kidneys – in terms of the properties and interchange of individual molecules – the mechanist's dream.

The grand unification between the concerns of the geneticists and those of the mechanist physiologists came in the 1950s, the 'crowning triumph' of twentieth century biology: the elucidation of the genetic code. This required a theoretical addition to the mechanistic programme, to be sure. Hitherto it had been sufficient to claim that a full accounting for the biological universe and the human condition was possible by an understanding of the trio of *composition*, the molecules which the organism contains; *structure*, the ways in which these molecules are arranged in space; and *dynamics*, the chemical interchanges among the molecules. To this now needed to be added a fourth concept, that of *information*.

The concept of information itself has an interesting history, arising as it did from attempts during the 1939–45 war to devise guided missile systems, and through the 1950s and 1960s in laying the theoretical infrastructure for the computer and electronics industries. The understanding that one could view systems and their actions in terms not merely of matter and the energy flow through them but in terms of information exchanges, that molecular structures could convey instructions or information one to the other, shook up a theoretical kaleidoscope. In one sense this made possible Crick, Watson, and Wilkins's recognition that the double helical structure of the DNA molecule could also carry genetic instructions across the generations. Molecules, the energetic interchanges between them *and* the information they carried, provided the mechanist's ultimate triumph, expressed as already mentioned in Crick's deliberate formulation of what he called the 'central dogma' of the new molecular biology:

DNA → RNA → protein

In other words, there is a one-way flow of information between these molecules, a flow which gives historical and ontological primacy to the hereditary molecule. It is this which underlies the sociobiologists' 'selfish gene' argument that, after all, the organism is merely DNA's way of making another DNA molecule, that everything, in a preformationist sense that runs like a chain through several centuries of reductionism, is in the gene.

It is hard to overemphasise the ideological organising function fulfilled by this type of formulation of the mechanics of the transcription of DNA into protein. The imagery of the biochemistry of the cell, long before Crick, had been that of the factory, a factory whose functions were specialised for the conversion of energy into particular products and with its own part to play in the economy of the organism as a whole. Some ten years earlier than Crick's formulation, Fritz Lipmann, discoverer of one of the key molecules engaged in energy exchange within the body, ATP, formulated his central metaphor in almost pre-Keynesian economic terms: ATP was the body's energy currency. Produced in particular cellular regions, it was placed in an 'energy bank' in which it was maintained in two forms, those of 'current account' and 'deposit account'. Ultimately, the cell's and the body's energy books must balance by an appropriate mix of monetary and fiscal policies.

Crick's new metaphor was more appropriate to the sophisticated economics of the 1960s in which considerations of production were diminishing relative to those of its control and management. It was to this new world that information theory with its control cycles, feedback and feedforward loops, and regulatory mechanism was so appropriate, and it is in this new way that the molecular biologists conceive of the cell – an assembly-line factory in which the DNA blueprints are interpreted and raw materials fabricated to produce the protein end-products in response to a series of regulated requirements. Read any introductory textbook to the new molecular biology and you will find these metaphors as a central part of the description of cells. Even the drawings of the protein synthesis sequence itself are often deliberately laid out in 'assembly-line' style. The metaphor does not only dominate the teaching of the new biology: it and language derived from it are key features of the way molecular biologists themselves conceive of and describe their own experimental programmes.

And not merely molecular biologists. The synthesis of physiology and genetics provided by an information theory containing a double helix was steadily extended upward from individuals to populations and their origins by biologically determinist writings like those of E.O. Wilson[2] and Richard Dawkins[3] (see chapter 6). These writings draw explicitly on molecular biology's central dogma to define their commitment to the claim that the gene is ontologically prior to the individual, and the individual to society.

For the mechanical materialists the grand programme which was begun by Descartes has now in its broad outline been completed; all that remains is the filling in of details. Even for the workings of so complex a system as the human brain and consciousness, the end is in sight. An immense amount is known about the chemical composition and cellular structure of the brain, about the electrical properties of its individual units, and indeed of great masses of brain tissue functioning in harmony. Neurobiologists claim to have shown how the analyser cells of the visual system, or the withdrawal reflex of a slug given an electric shock, are wired up, and to have found regions of the brain whose function is concerned with anger, fear, hunger, sexual appetite or sleep. The mechanist's claims here are clear. In the nineteenth century, Darwin's supporter, T. H. Huxley, dismissed the mind as no more than the whistle of the steam train, an irrelevant spin-off of physiological function. Pavlov, in discovering the conditioned reflex, believed he had the key to the reduction of psychology to physiology, and one brand of reductionism has followed him. For this tradition, molecules and cellular activity cause behaviour, and as genes cause molecules, the chain which runs from particular unusual genes, say, to criminal violence and schizophrenia, is unbroken.

So what's wrong with reductionism?

We have seen how reductionism emerged in intimate relationship with the rise of capitalism. There is also an implication that this is an inadequate, painfully blinkered way of viewing the world, despite the grandiose claims for universal synthesis offered by its ideologues, from Crick to Dawkins and Wilson.

Why won't reductionism work? We can identify a series of major flaws in reductionist thought. It is sometimes claimed that opponents of reductionism object to it only on political grounds, or as applied to humans but not to other animals – as if Descartes was alive and well amongst us. Whilst the political positions derived from reductionism, so in accord with the ideology of the new right in politics – from Thatcher in England and Reagan in the USA to la Nouvelle Droit in France and the writings of Italian fascism – are indeed the enemy, reductionist philosophies need to be confronted on their own terrain, irrespective of the politics they succour or espouse. If reductionism as a method of explaining the world is flawed, it is flawed for other animals as well as humans.

Reductionism begins by an assertion of ontological priority. That is, it claims that the individual is ontologically prior to the society of which that individual is a member, and the atom is ontologically prior to the organism. If we go back to our example of the frog muscle twitching, it insists either that *first* actin and myosin molecules interdigitate and *then* the muscle twitches, or that in some sense the explanation of the twitch in terms of molecules is more fundamental then any other explanation, the only 'true' explanation, the goal of 'real science'. Reductionism thus cannot accept that phenomena are *simultaneously* both individual and part of a greater unity. Reductionism in the sociobiological sense begins from this philosophical premise of the ontological priority of the individual over society. In the biochemical sense, it insists on the priority of the molecule over the organism.

Secondly, reductionism operates by a phenomenon of arbitrary agglomeration and reification. Consider the way in which reductionism uses words like aggression, in which it lumps together social phenomena such as war, strikes, discord between the sexes, football hooliganism, the space race, or whatever, as if they were all expressions of one unitary phenomenon labelled aggression. What is more, the phenomenon labelled as 'aggression' in society is simply seen as the sum of the aggressive properties of the individual members of that society. One sees how this works in the context of the various medical interventive strategies offered to cure 'aggression' in society, notably the surgical interventions proposed after the inner-city riots in the USA from the 1960s onwards (see chapter 2). Such strategies claimed that the explanation for such social phenomena as 'aggression' must lie within the individuals involved in the aggressive activity. Therefore one should intervene, and 'cure' the riots by performing appropriate surgical operations on the brains of selected ghetto ringleaders to remove the 'aggressive centres' within them.

The process has been first to assert the ontological priority of the individual and then arbitrarily agglomerise distinct phenomena. The third step is to say that if this 'aggression' (or whatever) is a property of the individual, it must be located 'somewhere' in the body. There is a powerful homunculus model which operates through reductionist thinking, so that properties of the organism have to have a localisation in the brain; there must be a *site* in the brain for intelligence, a *site* in the brain for aggression, for sexuality, and so on. Yet point localisation of individual behaviour within a set of cells or molecules within the brain, is a fallacious way of understanding how brains – or individuals – actually operate. Behaviour is an expression of the properties of the system, of the organism; it is not located in any one part thereof. Consider a lecturer speaking into a microphone, which gives, instead of amplified speech, a howl. Where is the howl located? Not in the amplifier, or the microphone or loudspeakers, but in a resonating feedback loop between them all.

The next step that reductionism takes (and it is a very powerful trend within the history of Western science) is a process that can be referred to as arbitrary quantification.

This is the belief that it is possible to quantify any particular property, whether it is aggression or intelligence or whatever, and rank it upon some linear scale, so that one can state of any individual that he or she is *x* per cent more intelligent, or more aggressive, than another individual. This process of quantification expresses a naive belief in the power of algebra and physics to explain phenomena. There is an assumption that all properties are linear, and can be mapped on some sort of scale. Again, let us take aggression as an example. Here is an example of how 'aggression' has now been studied in humans by way of the analysis of aggression in animals. The method involves taking rats, putting them in small cages, dropping mice into the cage and noting how long it takes the rats to kill the mice. If one rat kills a mouse in 2 minutes as opposed to another which takes 4 minutes, then the first rat is twice as aggressive as the second rat.

In order to complete the scientisation of what is going on, this 'mouse-killing' behaviour on the part of the rat is published in scientific papers and grants are given for a study of 'muricidal activity' (which makes it a great deal more scientific!). The way is then open to extrapolate this finding into the human situation, so that if a drug is found which affects the rate at which rats kill mice, this can be regarded as an 'anti-aggression drug'. (Drugs of this sort have now been clinically tried on 'schizophrenics' and what are described as 'violent hospitalised criminals' in Strasbourg.) So the reductionist programme generates a range of technological outputs, techniques of knife, electrode and chemical for interfering with brains, and thus modifying behaviour.

But to return to the analysis of the flaws in reductionism. Arbitrary quantification is the ascription of numerical values to qualities which cannot be adequately encapsulated in this sort of way. There follows what might be described as arbitrary algebraicisation; that is the belief that, if one can ascribe a number to some property, one must be able to partition out that property into a proportion given by nature and a proportion given by nurture, with a minor additive term for interaction. This nature–nurture dichotomy is another major example of the way in which reductionist thinking forces one into wholly spurious problems; the dichotomy between interacting processes each of which modulates the other is turned by reductionism into an arbitrarily polarised contest between two static and separate, quantifiable forces.

Reductionism having emphasised that all human phenomena can be divided up into a category given by genes and a category given by environment, with very little scope for interaction between them, it is scarcely surprising that when the algebra is applied it turns out that everything, or a very large proportion of everything, is inherited, whether it is radicalism, introversion, aggression, intelligence, twins' propensity to answer questionnaires consistently or inconsistently, or the ability to learn French at school. Such examples, drawn from heritability studies carried out over the past few years, offer conclusions which have substantial social resonance and political significance, but are strictly empty of scientific content.

The final step in the sociobiological reductionist model is to argue that, if all the properties of societies are merely the properties of individual members of that society, and these properties are genetic and hence inherited, it should therefore be possible to find an adaptive evolutionary explanation for them. This gives rise to the whole pattern of adaptationist myths, which are discussed in chapter 6. Although sociobiology is the most important current example of this sort of global reductionist thinking, I shall not discuss it here.

But evolutionary myths for the origins of what are regarded as adaptive properties of individuals are very popular at present. Just one example is the argument that the present division of labour between the sexes in contemporary Western society is given by an adaptation which derives from the human gatherer-hunter past, in which men went out and caught big animals whereas women stayed home and nurtured the children, and therefore men got genes for greater spatio-temporal ability whereas women got genes for linguistic ability. Hence, we are told, men are executives and women are secretaries.

The final part of the process is the Beatrix Potter syndrome, the attempt to project human qualities onto animals and then see the existence of such animal behaviours as reinforcing one's expectations of the 'naturalness' of the human condition. When one finds in the animal world exactly what one expects from the human world, one can then translate that observation back to the human world again so as to confirm it's 'naturalness' – children are naughty because Peter Rabbit behaves like a human child. This process is not new – it was described as long ago as the nineteenth century by Engels in response to the social Darwinism of the period – but when we read of baboon harems, propaganda-making ants, prostitution in humming birds or gang rape in mallard ducks (all examples the sociobiologists offer us) we are irresistibly reminded of it.

The processes outlined in this chapter are key steps in the methodology of reductionism in biology and show why it is both seductive and fallacious. Each of the steps in the argument has to be challenged, not merely because each generates dangerous technologies or poor ways of thinking, but because each contributes to preventing an understanding of the complex reality of the biological world.

Alternatives to reductionism

Biological reductionism has developed, historically, until today it is the major mode of explanation in biology. I have tried to show how reductionist explanations do more than merely sustain the ideology of the dominant class, race and gender, and generate obnoxious technologies; but are also fundamentally flawed as ways of describing, interpreting and predicting the world. That is, reductionism is at best a partial, at worst a misleading and fallacious, way of viewing the world. Reductionism, which began as liberatory, has become oppressive, like capitalism itself and today it limits our understanding of the universe.

There is no space here to discuss in detail the alternatives to reductionism explored in the two collective books *Against Biological Determinism*[4] and *Towards a Liberatory Biology*,[5] and in *Not in Our Genes*,[6] but it is necessary to emphasise some central points. This is not an attempt to replace biological by social explanations; the task is to integrate our understanding of both. Nor is it adequate to rather weakly offer models of the world in which events are partitioned out into *x* per cent biological, *y* per cent social, or whatever. It will not do to say that so-and-so is mad because of the way they were brought up, whilst such-and-such is mad because they have genes for madness; such a dualism leads only to a morass. Instead we assert the simultaneity of phenomena; everything is at the same time biological *and* social, and there are a variety of different languages in which we can analyse phenomena. There is no mind-brain dichotomy for instance; minds and brains are two ways of describing the same set of phenomena, and neither description is ontologically reducible to the other.

The task of relating different discourses, different levels, is one of looking for correspondences, for the translation systems which would enable us to interpret biochemistry in terms of physiology, physiology in terms of behaviour, and vice versa, without implying that there is a unidirectional arrow of causation which runs from the one to the other. There are of course a great many problems about the relating of discourses in this way. But in conclusion I must refer to one other aspect of the non-reductionist biological synthesis, that is a consideration of the role of the organism in the context of its environment, because this is central both to the reductionist metaphor of the environment as the selector of genetic programmes, and to our non-reductionist alternative.

For reductionism, there is first the genetic programme of the organism; the environment offers certain challenges, and the organism by virtue of its genotype responds to them. But we can transcend this passive view of the organism by understanding that it is an active maker and translator of its own environment. Organisms select environments, they work on and transform environments, and thus change their own history. This is manifestly true in the case of humans. But consider a much simpler example. Put a unicellular organism in a tube of water with a drop of sugar solution at one end of it. The organism will move towards the sugar solution, from a non-sugar environment towards a sugar-rich environment; it will *select* its environment. Once there, it will absorb the sugar, it will convert it metabolically into a variety of products and spit them out again. It will thus take part of its environment into itself, and transform it into more of itself, and at the same time it will actively *transform* its environment, which will change as a result of its being there. Following the change the environment itself may become less favourable to the organism – it may become more acid, for example, as lactic acid is produced. The organism will them move out of that environment once more to another, more favourable one. This very simple example is very far from a reductionist model of the passive operation of environment on the organism. Instead, one has a very simple organism actively transforming, interpenetrating with, making over and transcending its environment. This way of looking at the world is quite different from the reductionist perspective, and generates different experiments, different mathematics, different theories, different explanations.

A non-reductionist (what could be called a dialectical) biology, offers more, however, than just the insistence on this active interpenetration of environment and organism; it also insists that events observed at any given moment have histories – they are not the frozen moments in time which reductionism offers. Let us return to the example of the frog muscle twitch. Reductionism offers a frozen moment of explanation, of the muscle *now*; but we also have to recognise that the muscle *now* exists in the context of all of its past history, a history which a non-reductionist biology gives us; that is, an evolutionary history, a developmental history, and a historically contingent history.

This is, of course, even more important when we move from the study of non-humans to the study of humans. Reductionism offers a fixed linear view of human nature. This is very clear in IQ theory: the intelligence of a child is seen, as it were, as a small percentage of the intelligence of the adult, and development consists merely of the child moving linearly along some scale, until it reaches the adult size. Similarly, intelligence is seen as historically fixed. The intelligence of a child *now* is regarded as identical in quality to the intelligence of a Victorian or a feudal child.

A peculiar controversy has blown up recently over the discovery that the average Japanese IQ has increased significantly over the last several decades. There is considerable concern in the IQ literature as to how to account for this, as it manifestly cannot fit a genetic model. But once we recognise that intelligence is historically contingent, that whatever human nature is *now*, it is not the same as human nature was a hundred years ago, or five hundred, or a thousand years ago, the problems disappear. That is, properties such as intelligence, which are ascribed by reductionism to the individual, are in actuality properties which are aspects of the individual in a social context. The intelligent child 100 years ago might have been able to operate a multiplication table; the intelligent child today has at his or her disposal, as a result of the collective intelligence of the population as a whole, vastly more calculating power than was available to all the mathematicians from pre-historic humanity through to the pre-computer era. This has qualitatively changed the nature of intelligence, and humanity's relationship with both the natural and the social world. It is the constant transcendence of our own nature by ourselves which is the unique human biological quality. It is this which forms the problematic matrix for explanation by a unified biological and social science.

2. Neuroscience: the cutting edge of biology?

Seán Murphy

Towards the end of 1981, a year in which the Nobel Prize for Medicine was awarded to three neuroscientists working in the USA, the influential scientific weekly *Nature*[1] published a survey of the neurosciences. This survey claimed to be an account of the position of the neurosciences in the opinion of outside observers.
It began:

> In one sense, the neurosciences are at the cutting edge of the new biology . . . they are engaged with one of the most teasing – and oldest – questions: How does the physical representation of the mind, the brain, carry out its functions?

Answers to fundamental questions about the workings of the mind and brain, and indeed the relationship between them, have been sought for over two millenia by both natural scientists and philosophers. Today, that oldest question remains a teasing one. Although an understanding of the *physical* working of the brain will not necessarily reveal all about the mind, that has been the drift of research in the area over the last fifty years. Changes in technology, and particularly the ascendancy of molecular biology and particle physics, have driven workers in the neurosciences to look at ever smaller, isolated parts of the nervous system in their quest for answers to highly specific questions, often forgetting that the nervous system expresses its functions not in isolation but through continuous transaction with the world outside.

In part recognition of this fact, neuroscience now has practitioners in most branches of academic endeavour. 'Neuro' has become a ubiquitous prefix for psychologists, physiologists, morphologists, chemists, ethologists, immunologists, to name but a few. By gathering together groups of people from many disciplines it is hoped that apparently disparate aspects of nervous system function will become clearer at a holistic level rather than remain an unconnected string of experimental findings lost in the reductionist literature. Of course, there is the problem of communication between these disciplines. The technology that has been built up in the neurosciences is sophisticated and each discipline maintains its own language – it is an area where one 'specialises' very young. There is also the problem of diversity of interests: if honest about their motives, workers in the neurosciences would probably come up with one of two basic reasons for being there. There are those with the big, burning questions about nervous system functioning to answer. There are others who find the nervous system a convenient or interesting tissue with which to develop an idea or to test a broad cellular hypothesis. This is possible because of certain properties of nerve cells. For example, there is a tremendous diversity of cell size and shape in the nervous system. Nerve cells make very complex connections one with another, even though they have undergone one or more migrations during early development. For the molecular biologists, neurons express more of the DNA (produce a greater variety of proteins) than any other cell in the organism. There is no criticism intended of these varied motives. Either route of entry is absolutely permissible and both have added considerably to our understanding of the nervous system.

So what is it that attracts such endeavour, and much finance, to neuroscientific research? Certainly an understanding of nervous system functioning promises much – a 'solution' to the mind-body problem, an 'explanation' of language, the means of expanding mental achievement. But the obverse of that coin is that much is demanded from the neurosciences – 'cures' for mental illness, ways to restore function to limbs made useless by transected spines, justification for particular individual or social behaviours and biological reasons for social constructs. As Roger Sperry, one of the three 1981 Nobel prize winners put it:

> Ideologies, philosophies, religious doctrines, world models, value systems and the like will stand or fall depending on the kinds of answers that brain research reveals. It all comes together in the brain.[2]

Of course, with so much depending on the answers, it will come as no surprise that particular avenues of neuroscience research are often directed by those with different vested interest (that is, government departments rather than society as a whole). Consequently, the interpretation of results, and the uses to which these are put, expand rapidly from the chemical into the social realm.

In the same article,[3] Sperry goes on to suggest how:

> According to our [that is, his] latest mind–brain theory and its implications, Marxist-Communist doctrine is founded on some basic errors in the interpretation of science and of what science stands for and implies in reference to human nature and to social and worldview perspectives. As a result, the kinds of values upheld in Marxist doctrine are almost the diametric opposite from those which emerge from a scientific approach on our present terms.

Note, there is something called 'a scientific approach' and this, taken off the shelf marked 'guaranteed to be objective', and sealed in a vacuum, will reveal the whole truth. Sperry goes on to indulge his 'scientific approach':

> One of the best refutations of Marxism itself is that is was not Marx's actions in satisfying his material needs for subsistence that changed the world, but his philosophy, visionary ideas and Communist ideology.

Marx, according to Sperry, arrived in the world outside the womb with a visionary experience, presumably genetically endowed, thus beating the record previously held by St Paul and without environmental aids such as a blinding light or a road to anywhere. But it so happened that Marx experienced the vagaries of life in nineteenth-century Europe, the revolutions of 1848, was hounded from one country to another and spent most of his life in poverty at the wrong end of English industrialised society. From this continual *transaction* with his environment emerged Marx's interpretation of society[4].

Neuroscience does have much to contribute to significant clinical problems of particular societies, but there are dangers inherent in such an expectation. First is the assumption that the problems of a society are generated by particular individuals (manifesting their biologically determined behaviour) rather than by the society itself. Second is that the behaviour of such problem-individuals can be transformed into something socially acceptable by changing particular aspects of nervous system functioning. The belief, and its transformation into clinical practice, is illustrated in this chapter by two examples: psychosurgery and psychopharmacology. There are many reasons for selecting these two fields. First, they reveal the dangers of all too readily adopting 'causal' explanations. Second, psychopharmacology is probably the most frequently offered treatment in any GP's surgery. Third, the debate around such treatments has spread into the wider social sphere. And last, they highlight how the reductionist approach to answering questions about the nervous system can lead to full-blown reductionism.

The difference between reductionism and the reductionist approach is discussed throughout this book (see especially chapter 1). The reductionist approach is entrenched within the neurosciences. Faced with a system such as the brain containing more than ten thousand million (10^{10}) neurons, which between them make 10^{14} synapses or connections, then the sorts of questions one can ask of the *whole* interconnected and functioning system are somewhat limited. Over the last twenty years techniques have been developed such that now individual neurons can be studied either *in situ* or in isolation. The problem is that one cannot extrapolate from the properties of a single neuron to the system as a whole. In the liver, the properties of one cell are very much like that of another. In the brain, while most neurons appear to possess a basic set of properties which are not unlike those of other cells, these properties are modified, amplified or suppressed by interactions with other cells.

This more than anything is the unique feature of the nervous system – the elaborate intercellular connections. It is not only the fact that neurons demonstrate the ability to pass information from one to another, for many if not all cells are coupled to others in some way. But the nature of this coupling – in space and in time – is what is interesting. Recognising this organisation of cells and the connections within the brain, is it then possible to divide the brain into functionally distinct areas?

To a certain extent it is. For example, regions of the large cerebral hemispheres can be designated as being involved with processing sensory information or with initiating motor behaviour. Centres concerned with incoming sound can be distinguished from centres concerned with formulating words for speech. However, it is not correct to assume that these centres are in identical positions in every one of our species. In addition, it seems that the brain is not preprogrammed and that, especially in early life, if one area is damaged, an adjacent area can compensate and assume an additional function. Finally, the nervous system is not like a telephone exchange where you can close down one line, leaving all others unaffected. Invasion into one area will have repercussions in other areas of the system.

Psychosurgery

Individuals who have suffered brain damage (as through a stroke or tumour) are often changed in their behaviour. Certainly there may be partial paralysis and/or speech defects but accompanying these may be rapid changes in mood, unexpected violence or extreme reticence. Some of these latter changes may be expressions of the frustration of paralysis and inability to communicate, but others may be correlated with the site of the damage.

The 'classic' description of behavioural changes following damage to the brain comes from the story of Phineas Gage. In 1848, Gage was working on a new railway line in Vermont, USA. On the afternoon of 13 September, while blasting rock, there was an accident that resulted in a metre-long iron bar passing through Gage's left eye orbit and out through the top of his skull, clean through the front part of his brain. Miraculously, within minutes he regained consciousness and spoke to the gathered crowd. Gage survived this accident and lived on for many years. Before his accident, Gage had been a mild-mannered and responsible man, but afterwards his behaviour changed. He was described as a loud-mouthed obnoxious soul who cursed continually, exhibited little sense of purpose and wandered aimlessly across the country. Apparently Gage's intelligence was not substantially impaired but he lost the capacity to plan ahead and to care about the impression that he made on others. The personality changes perceived by his friends were so great that they declared 'he was no longer Gage'.

The frontal lobes of the brain, the site of the damage in Gage's case, make extensive interconnections with other structures deeper in the brain that comprise the limbic system. This system is implicated in the control of emotional responses. If damage to the brain results in changes in individual behaviour, then it was argued that brain manipulation could be employed as a means of behavioural and, therefore, of social control. The 1960s was, for the USA and a number of countries in Europe, a decade of social unrest as evidenced by hostile demonstrations, often against the 'authorities'. The war in Indochina was going badly for the USA and consuming vast amounts of money and American lives. There were riots on college campuses and on inner-city streets. John and Robert Kennedy, Martin Luther King and Malcolm X, amongst others, were assassinated. In 1967, in a letter to a leading medical journal, three Harvard physicians, Vernon Mark, Frank Ervin and William Sweet[5] suggested that mass neurological screening programmes be established to locate people with 'low violence thesholds' allegedly traceable to 'episodic dyscontrol syndromes' or a variety of 'limbic brain disease'. The argument was clear:

> That poverty, unemployment, slum housing and inadequate education underlie the nation's urban riots is well known, but the obviousness of these

> causes may have blinded us to the more subtle role of other possible factors, including brain dysfunction, in the rioters who engaged in arson, sniping and physical assault.
>
> It is important to realise that only a small number of the millions of slum dwellers have taken part in the riots, and that only a subfraction of these rioters have indulged in arson, sniping and assault. Yet, if slum conditions alone determined and initiated riots, why are the vast majority of slum dwellers able to resist the temptation of unrestrained violence? Is there something peculiar about the violent slum dweller that differentiates him from his peaceful neighbour?
>
> There is evidence from several sources . . . that brain dysfunction related to focal lesion plays a significant role in the violent and assaultive behaviour of thoroughly studied patients. Individuals with electroencephalographic abnormalities in the temporal region have been found to have a much greater frequency of behavioural abnormalities (such as poor impulse control, assaultiveness, psychosis) than is present in people with a normal brain-wave pattern.

Shortly thereafter, Mark and Ervin received substantial research grants from the US Law Enforcement Agency. Their particular philosophy was expounded at length in their book *Violence and the Brain*[6]. Whatever the original causes of the brain dysfunction, its damage was deep and irreversible.

> If environmental conditions are wrong at the important time, then the resulting anatomical maldevelopment *is irreversible*, even though the environmental conditions may later be corrected . . . The kind of violent behaviour related to brain malfunction may have its origins in the environment, but once the brain structure has been permanently affected, the violent behaviour can no longer be modified by manipulating psychological or

> social influences. Hoping to rehabilitate such a violent individual through psychotherapy or education or to improve his character by sending him to jail or giving him love and understanding – all these methods are irrelevant and will not work. It is the brain malfunction itself that must be dealt with and only if this is recognised is there any chance of changing behaviour.

Sweet, in the introduction to Mark and Ervin's book, states that 'human behaviour, including violent assaultive action, is an expression of the functioning brain'. This statement is fine if we are to understand by it that in order for a person to engage in violent assaultive action they must have a functioning brain. But as Stephen Chorover points out[7] in one of his many critiques of the practice of psychosurgery:

> In order to serve as a logical justification for psychosurgery it must *also* mean that violent assaultive action results *primarily* from (i.e. is *caused* by) something taking place in the brain of the person who engages in the action.

The psychosurgeon's argument is that there is something that can be located and treated within the brain of certain patients.

Public discussion of the case of psychosurgery came to a head in the late 1970s with a malpractice trial in Boston. Drs Mark and Ervin had allegedly been negligent in their treatment of one Leonard Kille. Kille was a young engineer who in the 1950s had suffered a ruptured peptic ulcer and had gone into deep shock and coma. After this incident his behaviour became 'unpredictable and psychotic'. He became paranoid and harboured grudges which eventually produced an explosion of anger. Innocuous remarks made by his wife triggered assaults. He continually accused her of 'infidelity with a neighbour'. A psychiatrist diagnosed temporal lobe seizure and referred him to Drs Mark and Ervin. They decided to treat his 'illness' by implanting electrodes in his limbic system. Electrical stimulation produced feelings of 'hyper-relaxation' in Kille. As these lasted for a few hours only, Mark and Ervin stated that it was obviously impractical to stimulate Kille's brain for the rest of his life. Instead, as a means of controlling the 'wild and unmanageable behaviour' they suggested to him that destructive lesions be made in the limbic system. Four years later (1970) Kille had not had, according to Mark and Ervin, a single episode of rage but continued to have occasional epileptic seizures with periods of confusion and disordered thinking. According to Kille's mother, who brought the action, this treatment rendered Kille of 'unsound and unbalanced mind and incompetent to handle personal affairs'. The jury in the malpractice trial cleared Mark and Ervin of negligence but declared Kille to be insane.

Mark and Ervin's diagnosis was that Kille's problem *originated* in his brain; not that his environment might have been involved. As the evidence showed, Kille was never observed to have a seizure or a fit of anger *away* from his wife prior to the operation. As it happened, at the time of the operation Kille's wife filed for divorce and proceeded to marry the lodger!

In August 1976 the National Commission for the Protection of Human Subjects of Biomedical and Behavioural Research issued its first report on psychosurgery. This commission had been created by the National Research Act (1974) to,

> conduct an investigation into the use of psychosurgery in the USA in order to determine the circumstances, if any, under which its continued use might be appropriate.

Surprisingly, given the events and public feeling that led to the setting up of the commission, this report fell only slightly short of *endorsing* psychosurgery as legitimate procedure,

> that can be of significant theraputic value in the treatment of certain disorders or in the relief of certain symptoms – a potentially beneficial therapy.

However, as Stephan Chorover points out in a critique of the two retrospective studies[8] that the commission considered, in one (at Boston University) the sample size was small and the variety of treatments large, and in the other (Massachusetts Institute of Technology) the follow-up study after treatment was too brief to permit definitive conclusions.

In some respects the era of psychosurgery is now over. Its passing is not necessarily the result of enlightenment but more because of the rapid advances in our knowledge of the ways that nerve cells speak to each other chemically. Psychosurgery is also expensive, involving pre- and post-operative care. There are also the risks associated with invasive surgery (and ensuing lawsuits). Psychosurgery has given way to psychopharmacology.

Before exploring the assumptions underlying drug therapy let us be clear about one thing. While we might reject vociferously the tenet that social order can be maintained by excising parts of 'disordered' brains, should we oppose all surgical intervention in the central nervous system? The answer must be no. A great number of people suffer from intractable pain, usually as a result of tumour removal or limb amputation. For the latter the analgesia produced by, for example, morphine, with the attendant problems of tolerance and dependancy, is not an adequate treatment. In some cases stimulation of nerve fibres in the spinal cord or in the brain stem, through permanently implanted electrodes, brings relief lasting weeks and even months.

Often it is not clear how electrical stimulation produces such effects. The best example of this is electroconvulsive therapy (ECT). Originating in the 1930s, ECT has been tried on both schizophrenics, and depressives, and is still widely used today. To an observer it appears a bizarre and incongruous treatment to intentionally generate convulsions with the use of an electric current applied across the brain. Ken Kesey[9] captures the paradox:

> Harding leans back to look at the door. 'That's the Shock Shop. Those fortunate souls in there are being given a free trip to the Moon. No, on second thought, it isn't completely free. You pay for the service with brain cells instead of money and everyone has

simply billions of brain cells on deposit. They won't miss a few.'

McMurphy listens for a moment. 'What they do is take some bird in there and shoot *electricity* through his skull?'

'That's a concise way of putting it.'

'What the hell *for*?'

'Why, for the patient's good, of course.'

'What a life,' Sefelt moans. 'Give some of us pills to stop a fit, give the rest shock to start one.'

The use of ECT is a clear case of adopting a general strategy to deal with a number of specific behavioural disorders, equivalent in psychological terms to the short, sharp shock so much favoured by law enforcement agencies. The results from ECT are very mixed and are often accompanied by feelings of disorientation and by amnesia. Despite these effects, ECT therapy is practised by psychiatrists across the world for the treatment of 'depressed' patients.

Psychopharmacology

For centuries individuals and whole societies have used mood- and perception-modifying drugs like nicotine, caffeine, alcohol and opium. Over the last sixty years it has become clearer how these drugs work and thereby novel agents have been created. When Otto Loewi first demonstrated in 1921 that the rate of a frog's heartbeat is modified by chemical substances (transmitters) released from the endings of the nerves that invade cardiac muscle, he achieved, simultaneously, two things. First, he provided evidence to support the belief that connections between nerve cells, or between nerve cells and their targets, are not continuous; that the nerve impulse is transformed from an electrical into a chemical signal at the junction between neurons, the synapse. Second, he opened the way for synthetic chemists and pharmacologists to set about designing and testing compounds that promoted or interfered with transmission across the synapse.

There are something like ten 'recognised' transmitter

compounds in the nervous system (that is, compounds that transmit signals between neurons). In addition, because of the complexity of the system, numerous other compounds influence the functioning of the nervous system. The distribution of transmitter compounds in the brain is fairly well known. For example, there are clusters of nerve cells in the mid-brain that produce the compounds serotonin, and other clusters, noradrenalin. The processes of these cells course through to the forebrain and make connections with other cells in the cerebral cortex.

Most of the psychoactive drugs (the psychotropics) influence the process of synaptic transmission. They do this in a number of ways: blocking synthesis or release of the chemical transmitter; speeding up or slowing down the breakdown of the transmitter; making the target cell insensitive to the transmitter; or simply by mimicking the action of an endogenous transmitter. For example, Largactil (a major tranquilliser) is used in the treatment of schizophrenia and it blocks the actions of the transmitter dopamine; Valium (a minor tranquilliser) modifies the sensitivity of cells to the transmitter GABA. The diversity and availability of psychotropics in Western societies has now reached staggering proportions. In 1979, over fifty million prescriptions were written in the UK for psychoactive drugs. The vast majority of these were for minor tranquillisers like Valium, Librium and Mogadon.

The high prescription rate is easy to understand. Psychotropics are reasonably cheap to make, easy to dispense and simple to take. Consultation time is kept to a minimum, there are no demands made on hospital facilities and, while there may be dangers of overdose and dependency, the effects of the drug are reversible. But the assumption underlying the prescription is the same as that underlying electrical stimulation treatment or psychosurgery – that there is an endogenous condition which is *causing* the depression or psychosis. This is not an argument that there are not *neurological* conditions in which one can identify abnormalities in brain form and function, and that drug therapy has no part in their treatment. A clear illustration of the value of drug therapy is in the condition known as Parkinsonism. The symptoms of this condition are fine tremour in the hands, difficulty in initiating movement and lack of motor co-ordination. There appears to be a deficit in a particular transmitter (dopamine) in one of the brain pathways concerned with muscle control and a certain amount of remission can be gained by Parkinson sufferers if they take L-dopa which is involved in the synthesis of dopamine.

Interestingly, people taking large quantities of L-dopa can display side effects including hallucinations. The similarity between people on high doses of L-dopa and the behaviour of schizophrenics led, in the 1970s, to the idea that the cause of schizophrenia is too high a level of dopamine in particular brain regions or that the neurons in these regions are oversensitive to dopamine. A functional (psychological) disorder is now suddenly equated with a neurological disorder and the treatment prescribed is a dopamine-blocker like chlorpromazine (Largactil).

That certain compounds have psychotropic effects, that these compounds interfere with the function of the nervous system in some way and that the administration of particular compounds to people with particular functional disorders alleviates their conditions, are all without question. What is questionable is that there is any causal link between these observations. But there are arguments which run: 'better to treat the symptoms even though the causality is unknown because you can die of the symptoms', or even: 'drug therapy has taken people out of institutions and back into the community'. The problems with these arguments are that it may be the community that is inducing the condition and that drugs are simply a way of adjusting to the social environment.

So how is it that millions of prescriptions for benzodiazepines are filled each year, and how do thousands of GPs come to prescribe these drugs? Anxious patients constitute over 30 per cent of all people seen in the surgery. Anxiety neurosis is defined clinically as:

> A disorder in which the principal manifestation is
> excessive anxiety, often amounting to panic,
> presenting in the psychic and/or somatic field.

It is a level of anxiety that interferes unacceptably with the patient's work, social life and view of herself. (The use of 'herself' is not gratuitous but reflects the fact that women of 40–60 years of age represent one of the largest patient categories. No doubt this results both from the willingness of GPs to prescribe such drugs to women and from their perceived need for them).

As the clinician sees it, the options are drug therapy or an alternative such as a form of psychotherapy or relaxation therapy. GPs have been criticised for slavishly using a biomedical model which encourages a view of the patient's distress as organic and suitable for medical resolution. It is also suspected that they succumb to work pressures in substituting drugs for time.[10] Patients may want to perceive their condition as organic. It is 'easier' to take the doctor's pill than to proceed with any process of self-examination. This leads to a cycle of reinforcement whereby the patient presents with the expectation of prescription and the doctor, for the reasons given earlier, fulfills that wish. The cycle must be broken for two reasons. First, the myth will otherwise prevail that a reasonable existence can be supported only with chemical adjusters. Second, persistent use of such drugs brings with it problems of dependence, and a new cycle is begun. In a recent survey[11] it is estimated that 100,000 people are dependent on minor tranquillisers in the UK.

Attitudes of younger GPs to the prescribing of tranquillisers provides some hope. Young doctors, with their training just behind them, may be more reluctant to prescribe. We can but hope, as far as tranquillisers are concerned, that they do not become more orientated to the demands of their patients as their experience in general practice grows. The dramatic increase in the prescribing of drugs is again taken up in chapter 4.

Endpoint – the direction of neurobiology

Naive adherence to the 'scientific approach' has led neurobiology to assumptions about the causality of human behaviour. If the behaviour does not fit or cannot be predicted then an organic answer is sought and the condition is 'treated'. In the same way that psychosurgery was attacked in the 1970s, so too is the extensive practice of psychopharmacology in the 1980s. Its replacement with less invasive methods such as psychotherapy, a time consuming treatment, will require an expanding rather than the contracting health service that is at present found in many countries, particularly the UK.

This chapter has focused on one particular aspect – the translation of a progressively reductionist interpretation of the working of the brain into clinical practice. The assumption of the psychosurgeons is that there is a distinct relationship between a specific brain region and a particular set of behaviours. In the same way, others have attempted to show that there are differences in structures lying in the left and the right halves of the brain and that this 'lateralisation' is responsible for differences in individual (particularly sex-related) behaviour patterns.[12] Instead of describing the sudden rush of anti-anxiety drugs, the rise of psychopharmacology could equally well have been illustrated by the proposed use of long-acting hormones (progestins) for the treatment of male sex-offenders[13] (see chapter 3 in this book).

However, recent advances in the neurosciences have furthered pain relief, the treatment of crippling diseases and provided means for the early diagnosis of neurological conditions. Neuroscience now stands on the brink of a description of how neurons become 'wired-up' and an understanding of these processes promises the means to stimulate new and viable connections between cells in damaged areas of the spinal cord.

How is it possible, then, to ensure the transfer of relevant neuroscience into the clinical setting and to guard against unsuitable treatments that are based upon false premises? Thus far, the job of watchdog has been left to a few disaffected clinicians and vocal neurologists. What is needed is the greater involvement of people, scientist and non-scientist alike, in the planning and practice of science policy. Until then, we will be forced to react and counteract rather than create.

3. The determined victim: women, hormones and biological determinism

Lynda Birke

This chapter focuses upon a specific field of biology and medicine, and looks at instances of ideas that human behaviour is directly determined by biology. Over the last century, an array of arguments have arisen which purport to explain features of human behaviour in terms of the levels of particular hormones carried in the blood. That something in the blood may be thought of as *causing* behaviour of some sort is hardly a new idea, of course: it was, for instance, implied in medieval notions of the humours, each of which was inextricably linked to a particular temperament. However, the discovery of the various different hormones has greatly facilitated the proliferation of such ideas, and we can now find suggestions of a hormonal basis for a variety of behaviours, including aggression, sexual preference, a predisposition to being a victim of violence, being good at typing, attempting suicide, and many more.

If anyone is asked what hormonal differences we might expect to find between different individuals, they are likely to think first of hormonal differences between the sexes. These are not absolute (that is, no hormone is produced exclusively by one sex and never by the other), but quantitative differences do exist. It is, then, not surprising that most of the ideas of hormonal determinism have been specifically related to sex rather than, say, to race, class, or any other way of classifying people. For this reason, this chapter is principally concerned with illustrating deterministic ideas which relate specifically to gender, and the extent to which our understanding of the significance of hormones is impoverished by this inherent reductionism.

We are concerned in this book with the social and political consequences of particular areas of biology, and in the field of hormones and behaviour there are two consequences of particular importance. First, attempts to explain human behaviour as natural and inevitable serve to oppose social change, simply because they provide justification for what already exists. If, say, someone produced a hypothesis that the rather low number of women in engineering was due to female hormones, then there is not much point in bemoaning the fact and discussing alternative forms of technical education. It is no accident that these kinds of ideas are frequently espoused by those opposed to any progressive social change.

That these ideas serve to legitimate the existing social structure is a feature in common with other areas discussed in this book. However, the second consequence of hormonal theories is perhaps more serious, that is, the application of the knowledge. It is quite clear that hormones have been, and are, utilised in a variety of contexts. The majority of these are probably considered beneficial: insulin, for example, keeps alive people who a century ago would have died from diabetes. No one would dispute the benefits that arise from such uses of hormones. However, there are many uses of hormones which are less clearly beneficial, and about which there has been some controversy. These are usually the uses which we might consider to be more overtly political, such as in implementing population policies (e.g. hormones as contraceptive agents) or in the attempt to control sexuality. We will consider some of these later.

While the stated objectives of such hormone use may seem at first sight desirable, this often hides an element of social control. This kind of control, moreover, does not operate randomly, but against *particular* groups of society, usually those which are socially disadvantaged. It is not, for instance, white middle-class women who are likely to be given injectable hormonal contraceptives in the UK; it is most often working-class, or immigrant women.[1] This raises many questions: who, for instance, actually benefits from a particular medical intervention? And in what way?

The discovery of hormones

While the effects of certain hormones have been known for centuries (for example, the effects of castration have long been used in the management of domestic animals), the actual discovery of the physiological role of the endocrine glands and their secretions has occurred only within the last century.[2]

Once the initial discovery of these effects had been made, the hunt was on to work out what the active agents produced by the glands were. The first few decades of this century then saw the isolation and purification of a range of hormones secreted by the endocrine glands, including several of the 'sex' hormones. To begin with, research into the relationships between these hormones and reproduction was painfully slow, largely because such research was held to be socially unacceptable. This was particularly true of the research which eventually led to the production of the contraceptive pill, and resulted in production being delayed for more than twenty years. This prejudice remained until the early post-war years, when the social and political climate changed sufficiently for the pharmaceutical companies to begin clinical trials (although largely on women in the Third World) and to market, in 1960, the first 'Pill'.[3]

In parallel with the development of research into hormonal physiology, there grew an interest in the relationship between hormones and behaviour, much of which was explicitly seen as part of a broad research approach to the solution of human social and sexual 'problems'. In the USA, for example, the National Research Council's Committee for Research into Problems of Sex was, as its name implies, primarily concerned with such 'problems', and sponsored a wide range of sex studies, both of humans and of other animals. From 1922 until after the Second World War, this committee funded an enormous quantity of research into 'problems of sex', covering a wide range of projects, from basic research into the physiology of gamete production and the relationships between hormones and sexual behaviour, to studies of human sexual behaviour, to studies of human sexual practice such as the Kinsey surveys.[4] Much funding for research into reproductive biology came from the Rockefeller Foundation, whose concern was with what they considered to be pressing social problems. Accordingly, the practical and ideological emphasis of such research was on the transformation of 'human sex into a scientific problem', thereby making it an 'object for medical therapy of all kinds of sexual "illness", most certainly including homosexuality and unhappy marriages'.[5]

A primary concern of this research, then, was the amelioration of *human* social problems. Thus, while much funding supported research into mechanisms of sexual behaviour of other mammalian species, there was, and there remains, a concern to relate the findings to human society, often by means of direct comparison between the two. The use of data derived from animal studies in relation to specifically human problems is, however, particularly controversial. One can frequently find animal data which either support or refute a fashionable hypothesis. The best that can be said is that animal models can provide us with a baseline of information, on which we can generate questions concerning human physiology or behaviour. And even this is fraught with difficulties. If a given drug is associated with cancer in some laboratory animals, but there are, to date, no data suggesting such an association in humans in clinical trials of the drug, can we assume that the drug is *not* likely to provoke cancer?

Moreover, the validity of animal data may be questioned at times simply because they *do* indicate a potentially toxic effect: when opponents to a particular drug point to the existence of a set of data which indicate noxious effects of the drug, there is often an uproar about the validity or otherwise of such animal data, often from the very companies which manufacture it. Yet if animal data provide *such* a poor model then (a) we should ask why the testing was carried out at all, and (b) we would realise that we would have to test all potential drugs on human guinea-pigs. Clearly, it is a working assumption of drug manufacturers and doctors engaged in clinical trials, that animal testing does indeed give some indication. That this is waived in some instances is essentially a political consideration, not a medical one.[6]

The history of research into reproductive biology, then, has been influenced in several ways by a concern with human social or sexual 'problems'. Given this, we should not be surprised to find a number of assumptions – derived from animal experiments – about the hormonal determination of human behaviour. There are some specific examples to illustrate some of these more general points. To begin with, we shall consider sexuality, and the 'sex' hormones, perhaps the first thing which feminists have criticised as an instance of biological determinism. The second area considered is completely different, dealing with the suggested relationship between hormones and being a victim of repeated violence. The third example is the idea of hormonal *deficiency* in older women, a deficiency which, we are assured, can only be cured by daily hormones.

Hormones, sexuality and 'sexual disorders'

The influence of steroid hormones, particularly the so-called sex hormones, on sexual behaviour and its development has been studied in several mammalian species, although most of the research has been on rodents, especially rats, and on some primate species such as rhesus monkeys. The significance of these hormones for the display of behaviour which we might regard as 'sexual' (e.g. the postures which permit copulation) varies between species. For some, such as the rat, removal of the source of hormones largely abolishes this behaviour. For primates, such a rhesus monkeys, the hormones are not as essential.

The burgeoning research into hormonal mechanisms underlying sexual behaviour in other mammals provided much information about the relationship between behaviour and endocrine secretion. It was, however, largely reductionist, focusing on a complex interaction between two individuals and relating limited features of this to variation in the levels of particular hormones in one of these individuals. Very little of this research has seriously considered the possibility that factors other than hormonal variation might be important, or that the social interactions themselves might change the levels of hormones being measured. What we know, then, about hormones and animal sex is firmly within the scientific tradition of reductionist explanation.

One research area in which, for example, explicit links are commonly drawn between animal studies and human behaviour is that on homosexuality. Since the early 1960s a great deal of research on hormones has focused on their effects during early life on the differentiation of the body into female-type or male-type. The details are not necessary here: suffice to say that steroids, particularly androgens and oestrogens,[7] strongly influence early development. In rats, for instance, the levels of these hormones during early life influence the brain such that some features of adult behaviour are affected. These hormonal effects seem to underlie some of the behavioural differences between the sexes. These findings have been used as the basis for the idea that if hormones influence the sexual behaviour of rats, then perhaps they are responsible for what are described as sexual deviations, particularly homosexuality. The equation is straightforward: a 'disordered' sexuality must be causally related to 'disordered' hormones.

In making these suggestions, researchers make a number of assumptions. First, they suppose that rats provide a plausible analogy for human sexual behaviour. This seems unlikely. Human sexual responses are immensely complex, and culturally variable. People in the West may find it hard to imagine how rubbing noses can turn you on sexually, yet that is a significant part of the sexual repertoire in some societies. Drawing parallels between this enormous variety of sexual expression and rat sex – which, as far as is known from research does not vary much from rat to rat – seems far fetched indeed.

Second, the legitimacy of the reduction from complex human behaviour to hormones is not questioned, but is simply assumed. Furthermore, it is assumed within a tradition which lumps together *all* forms of sex other than monogamous heterosexuality as deviant, obscuring the complexity of different phenomena by labelling them all the same. Within the literature on homosexuality, for instance, one can find studies which conflate data about homosexuals with those about transsexuals, with those about tranvestites, or with bisexuals. The hormonal bases of these enormously heterogeneous categories of 'deviant' sexual expression are then sought.[8] The classifications that we use to describe sexuality (as is the case with much else of human behaviour) are, however, vague; and they are vague precisely because human beings *are* so flexible that they cannot always be slotted into neat boxes. And, of course, if these boxes are vaguely defined, then the problems are compounded if we aggregate them.

More importantly, these classifications are frequently used as though they represent some unitary biological reality. It is not that biology is irrelevant to these classifications: biology clearly has something to do with the genitals we each possess, and is therefore relevant to the social meanings which we attach to our definitions of sex and gender, and thence to all the concepts which our society attaches to gender.[9] But that is a far cry from saying that each of these classifications, including gender itself, is absolutely synonymous with some definable biological state. It is simply not possible to map, say, 'homosexuality', or 'transsexualism'' on to biological causes.[10] One response to this incommensurability might be, and has been, to propose that each category be subdivided. Thus, one can speak of several 'homosexualities'[11] – some of which might then be referred back to a hypothetical biological substrate, while others are not. But fining the boxes down merely limits the range of the determinism; it doesn't challenge it. This might be called the liberal answer to the problems of determinism. Having agreed that there is a problem associated with the crude categories to which determinist arguments are attached, and that such crude determinism only serves to limit human potential, the proposal is simply to expand the categories. Pluralism, however, does not help us much.

Much of the work done on sexuality and sexual preference is, as has been hinted, bad science: low numbers of people studied, lack of adequate control groups, lack of any attempt to control for the possibility that the person is taking some drug, poor methods of assaying hormones, and so on. These detailed points have been extensively criticised elsewhere.[12]

What is more important to note, is that portraying different expressions of sexuality as a product of a deranged biology serves to perpetuate the idea that they *are* abnormal: it certainly does nothing to change the basic assumption that exclusive heterosexuality is normal, and therefore morally right. What is not asked is *why* heterosexuality should be so exclusively valued in our society, and why alternatives are still so stigmatised and punished. We should remember that these assumptions, upon which so much of the research rests, are specifically *Western* concepts. The permissible forms of gender role, and of sexual expression, vary markedly from culture to culture and, indeed, have varied historically in our own culture. True, all known societies make some distinction between behaviour considered appropriate for one sex rather than the other – but this distinction is by no means everywhere the same as ours.

What we can say however is that, poor science apart, much of the research on human sexual preference rests on very shaky foundations. Popular misconceptions and assumptions underlie the experiments which are carried out, and determine the interpretations given. There are rarely any attempts to assess the assumptions made within a particular piece of research; the most that might be done is to point to the possibility that any endocrine differences that are found between, say, homosexual and heterosexual groups mights not be directly causal, but may themselves be a product of differences in behaviour. But none of these medically-based studies do anything to challenge the ideology of sexual expression within the family, which is an ideology rooted within the dominant culture of Western society and insists upon the primacy and 'naturalness' of heterosexual, procreative and family-centred relationships. On the contrary, medical research aimed at locating endocrine bases to homosexuality is serving to maintain that very ideology.

Victims of violence or addiction to hormones

Part of the stereotypes of femininity and masculinity in our culture is the dichotomy between passivity and aggression/violence. That there are many individuals who do not conform to these stereotypes is obvious. However, the existence of the stereotypes, and their continual maintenance within our society (e.g. by media images, processes of education, discriminative practices, and so on) sets the conditions for a considerable social pressure to conform to those stereotypes.

As we have seen in the case of sexuality, one way in which the stereotypes are maintained is by the advocacy of theories rooted in biological determinism. The stereotypic trait is seen, not as idealised, but as actual, and putative biological bases proposed for it. With respect to aggression/masculinity, versus passive femininity, there are a variety of such arguments.

Mostly, such deterministic arguments concerning aggression are articulated in opposition to feminist ideas, although some avowedly feminist writers have written similarly deterministic ideas of male aggression.[13] Such arguments tend to take something which, though hard to define, is largely thought of as an attribute of individuals (aggression), and then to equate it with a general phenomenon at the level of society (male domination). Having made the equation, biological *causes* are then inferred. Steven Goldberg, for instance, in *The Inevitability of Patriarchy*[14] argues that (a) male domination of society is universal, and that (b) the cause must lie in the early 'hormonalisation' of the male brain. Social expectations or ideology are, it would seem, irrelevant.

However, arguments such as these are not the only kind of deterministic notion that centres on the aggressivity/passivity stereotype. A more recent suggestion is one concerning some of the women who end up in Women's Aid refuges. Such women have often suffered many years of violent assaults, both on themselves and on their children. Yet despite these appalling conditions, some of them at least, have been portrayed as being 'victims of hormones'. In 1981, Erin Pizzey and Jeff Shapiro published an article, entitled 'Choosing a Violent Relationship',[15] followed in 1982 by a book, *Prone to Violence*.[16] In it, they claim to describe women who repeatedly end up in refuges, apparently being rather 'prone' to being battered, or having 'chosen' to do so.

Now we might speculate on a variety of reasons why a

woman might find herself involved in a repetition of violence. She might, for instance, return to the man who previously assaulted her because she cannot afford anywhere else to live, because she believes he will change, because he will otherwise gain custody of her children, or for many other reasons. Pizzey and Shapiro, however, ignore these reasons; rather, they come down firmly on the side of biological determinism. These women repeat the experience of being battered because, it is claimed, they have become addicted to a hormonal surge accompanying the horrifying violence. They have become what Pizzey and Shapiro like to term 'cortisone personalities'.

Much of the 'biology' on which the arguments in the book rest is quite simply wrong. One of the most glaring mistakes is in the terminology of a 'cortisone' personality. *Even if* a whole personality could be described in terms of *one* of the hormones secreted by the adrenal cortex (and that is a dubious assumption indeed) then it is not likely to be corti*sone*, since corti*sol* is the principal adrenal hormone of its type in humans. Perhaps they had some other species in mind. There are other ways in which the suggestion of a determining role for hormones is simply wrong, such as the suggestion that they stimulate 'pleasure centres' in the brain, but these will not be elaborated upon here. The point is not only to recognise the inadequacies of this pseudoscience, but to emphasise again the ideological foundations of such ideas and to point to likely consequences of such determinism. Pizzey and Shapiro's answer to these cases of repeated violent assault is firmly biological. It can be sought, they suggest, in drugs aimed at correcting the 'imbalance'.[17]

The 'prone to violence' story omits any serious consideration of social factors; rather, biology is seen as primary. Pizzey refers to the early childhood of these women during which, she argues, they became hooked on the hormone; thereafter, it takes over, and the social context in which any future violence occurs is made largely irrelevant. But this is to deny the social relations between women and men in a society which actively condones acts of violence towards women, and which encourages women to acquiesce and remain passive. Those social relations inevitably structure the ways in which we respond to any given situation, including violent or dangerous ones. People often ask why so many women remain for long periods of time with a man who regularly beats them up. The answer lies in part in women's powerlessness. To remain, even in a dangerous situation, may at least be predictable; to leave too often means considerable uncertainty – nowhere to live, no job, risking losing custody of children, and trying to live on a pittance from the state. To ignore all this and to focus on the individual woman is simply to remove the blame from where it should lie. And biology has little to say about powerlessness.

Ageing women: hormones and the menopause

The menopause symbolises a time of great change for women, a change which is often distressing. Partly, this is because of fear of physical problems associated with the menopause (see below) and partly because, for women, the menopause has become symbolic of ageing. Partly, too, the menopause coincides with life changes, such as the change in life-style as children leave home. This distress might be experienced in a number of ways: women may feel useless as their children leave home, they may become depressed, they may worry excessively.

These consequences may, of course, be interpreted in a variety of ways, but here we are concerned with the increasingly popular view that they are somehow *biologically* caused, along with the hot flushes, drying of the skin and a whole host of other physiological changes. The culprit in all cases is usually assumed to be a lack of oestrogen. Now, it is undoubtedly true that oestrogen secretion by the post-menopausal ovary is considerably less than that secreted by the ovaries of a younger woman. It is also undoubtedly true that many women suffer considerable discomfort from some of the physical changes which accompany the menopause. However, the tendency to leap to conclusions of biological causation has had a number of consequences.

The first consequence is that, yet again, such explanations of changing experience focus on the individual, thus obscuring any consideration of social context. Yet, as outlined above, the menopause symbolises a time of great change for women, including a decline in status. That hormonal changes accompany this most certainly does not mean that they cause it. If we are to understand why women find the experience of menopause distressing, then we have to understand it in much broader social/political terms. Even the apparently 'physical' changes, such as hot flushes, may be better understood as a product of complex interaction between the biological changes and the societal ones. Many of the physical changes, including flushes, respond to some extent to placebos (sugar pills), suggesting more of a psychological than purely physical base. That is not to say that women 'imagine' their sufferings. They most certainly do not. However, the complex interaction between biological events and a changing experience of social role can only be experienced at the individual level. That this experiencing facilitates some bodily events does not amount to 'imagination'. Some advocates of the hormonal-deficiency view see the loss of oestrogen as causally linked to a loss of *femininity*. Thus, Wendy Cooper, an extreme advocate of this view writes:

> For [oestrogen] is the key to a woman's health and happiness, and it is at work in the body even before birth . . . In many ways oestrogen has a place in the life of a woman rather like that of love itself. She is born with both, and both assume their greatest importance in her fertile years.[18]

This view was also enthusiastically endorsed by a Dr Wilson, who, in 1966 published a book describing the 'deficiency disease', and how it might be treated by hormone injections. He entitled it *Feminine Forever*.

The major consequence of locating menopausal problems within a strictly biological framework is, of course, that it renders chemical treatments possible. Hormone Replacement Therapy (HRT) has now become quite common both in the UK and the USA, and is now promoted with considerable vigour (despite lawsuits pending) by the manufacturing drug companies. Marilyn Grossman and Pauline Bart cynically observe:

> A 'deficiency disease' was invented to serve a drug that could 'cure' it, despite the suspicion that the drug caused cancer in women. That the suspicion had been voiced for so many years before anyone chose to investigate it is yet another example of how unimportant the well-being of women is to the men who control research and the drug companies who fund much of it.[19]

Perhaps not surprisingly, HRT has come in for criticism. It is certainly true that oestrogen injection can alleviate some of the physical distress which can accompny the menopause. For this reason, even the most cautious doctors are prepared to use it to tide a woman over a short period of particularly distressing experiences. The objections, however, come with keeping women on hormones for long periods of time (in the case of one New York woman, for over thirty years), and to the massive propaganda campaign which has promoted HRT.

The first point to be made is that we really don't know what the long-term effects might be. There has been much debate about whether or not long-term HRT might cause or aggravate cancer – but the fact remains that we simply do not know. As Paula Weideger, in her book *Female Cycles*,[20] commented:

> It is not enough to say that every woman has a right and a privilege to choose her poisons and name her saviours. I want to choose my medicines *after* I know their side-effects, not after their side effects have ravaged my body.

The second point which can be made is that, scarcely surprisingly, the campaign to promote HRT is enthusiastically supported by several pharmaceutical companies, who might be expected to give support to a drug which might be taken daily over many years by a large proportion of the population. The link between these organisations and seemingly less 'political' organisations such as the Family Planning

Association (FPA) should not be forgotten either. At one FPA sponsored talk, the virtues of HRT were being extolled (and none of the possible risks given serious consideration) to a background of posters and stalls promoting the products of the different companies involved.

The third point is that the advocacy of HRT occurs firmly within the context of women's oppression, despite the 'liberated woman' image used to advertise the drugs to the medical profession.[21] The concept of eternal 'femininity', a central plank of the campaign, is one which conveys a stereotype which feminists reject. Rather than seeing a search for 'femininity' as part of our oppression, we are encouraged to think of it as something fundamental to our lives as women which we have to maintain at all costs. Hardly a liberating thought. Moreover, another feature of the advertising is that HRT can be seen as a way of fitting women back into the family. Wendy Cooper, for instance, stresses the ways in which the *family* might suffer if a woman allows herself to become obviously menopausal. The emphasis on family responsibility means that we can view HRT as a prophylactic enabling women to return to, or continue in, their present role. Rather than allowing women to see that a major source of their problems at the time of the menopause is actually social oppression, they are instead offered a technological fix. Wendy Cooper concludes her book with a defence of both (white middle-class) femininity *and* the technological fix. She says:

> And maybe when the words 'Women's Lib' [sic] are no more than the remembered echo of some old battle-cry from an ancient victory, the words 'Biological Lib' will continue to have meaning for women, as the basic freedom, not designed to deny or decry the female role, but *to control and maintain it* [emphasis added].[22]

Biology is clearly meant to be our destiny.

Conclusions

We have drawn on three somewhat disparate examples of

the ways in which complex human behaviour can be seen as removed from its social context and reduced to the operation of hormones. All three examples have in common that the arguments used are crudely reductionist, and that they serve to define people in terms of a deviant biology. In this sense, such arguments serve to legitimate the existing social order: if we are merely victims of raging hormones, then what hope is there of social change? Furthermore, some of these arguments lend themselves to forms of medical treatment based upon hormones. That the hormone 'treatment' will do little to change the situation is not relevant; that it is used without full knowledge of the long-term consequences of drug use is. Indeed, in the case of hormones prescribed for the menopausal woman, we can see that there is some basis for suggesting that the idea of a'deficiency' was specifically articulated in order to provide a 'disease' for oestrogens to 'cure'. That is not to say, of course, that oestrogen treatment does not *sometimes* benefit *some* women whose experience of the menopause is particularly traumatic. The problem lies with the overt manipulation of a potential market for HRT, one consequence of which is that large numbers of women now seek HRT from their doctors, even when there is little need for it. They have been sold the illusion of 'femininity forever', purchasable, so the illusion goes, with oestrogen therapy.

Having agreed that some of these arguments lend themselves to direct application in the form of medical treatment, many scientists then absolve themselves from responsibility for the consequences of their research by assuming that the science *they* do in the laboratory is somehow different, somehow less tainted by commercial interests.

There are several things which such a view ignores. First, it assumes that the scientist *can* be 'purely' objective, *can* stand back from the society of which s/he is part and look disinterestedly at nature. Second, it assumes that the scientist is a somewhat passive observer of nature, simply using apparatus or reagents to uncover nature's laws. This ignores the reality of science as a form of production of knowledge. In the way that the scientist generates hypotheses, what s/he considers to be data worthy of collection or relevant to the establishment of hypotheses, and the interpretations given to any data obtained – all of these mean that the scientist is an active agent in that production of knowledge. And all of them mean that scientists' social values become embodied within that knowledge as it is produced.

These considerations are as true of hormone research as they are of any other. It would be naive to suppose that endocrinology is somehow free of values, but that the drug companies and/or the medical profession are to blame for the way that they misuse, say, oestrogens. As we have seen, the very concepts which are used in research carry ideological weight. Having used concepts which themselves carry particular messages about human society, it is then but a small step to see human society as somehow reflected in the animal world. And once the analogy has been argued, of course, human behaviour can so often be portrayed as merely a product of some feature of biology, be it unbalanced hormones or reified genes.

In this chapter, I have used three specific instances of ideas of biological determinism to illustrate more general points. The choice is fairly arbitrary and there are many other examples which could be used within the field of hormone research. However, the examples chosen were intended to do two things. First, to show the relationship between the production of scientific knowledge and social/political values, a relationship which, in some instances, leads to specific forms of social control, such as hormonal drugs aimed at manipulating a person's behaviour. Second, to show that our understanding of *biology*, of the living organism, is impoverished by our insistence on reductionist explanations. We focus on hormonal variations and assume that they *cause* behavioural changes, rather than attempting to view the organism as engaged in a constant interaction with its environment. This limited focus helps to maintain certain views of human behaviour. But it is worth remembering that such an approach is not likely to give us much idea of the rat or the rhesus monkey either. Even a rat is more than just a bag of hormones.

4. Pharmacology: why drug prescription is on the increase

Lesley Rogers

Drugs and disease

We live at a time when drugs are taken in abundance in industrial societies. Many of these drugs are prescribed by doctors, many can be self-prescribed since they are available from chemist stores without prescription or at local grocery or liquor stores, while others are taken inadvertently in food (e.g. antibiotics, hormones, or caffeine) or in the water supply.[1] The last two categories are diverse and it is virtually impossible for an individual to assess his or her drug intake from these sources, particularly in countries like Australia where manufacturers of prepared foods are not legally bound to state the exact contents of their products on the package labels. Most packaged foods also contain considerable quantities of preservatives and the controversial taste-enhancer monosodium glutamate.[2] Even fresh foods contain unknown amounts of chemical sprays, especially if grown in large monocultures where herbicide and insecticide spraying is the accepted means of farming.[3]

Drugs are not just tablets or bottles of strange liquid prescribed by the doctor, nor are they just substances which some people take to give themselves mental 'kicks' (e.g. marihuana, LSD, or even coffee and alcohol). The term drug is difficult to define but it could be said to refer to any administered or ingested substance which is foreign to the substances naturally present in the human body (e.g. analgesics, antibiotics), or any biological molecule taken by an individual in amounts in excess of those naturally occurring in that individual's biological system (e.g. oestrogen, insulin, monosodium glutamate). This chapter will be mainly concerned with drugs prescribed by the doctor.

In the USA there has been a steady increase from

1950–77 in the variety of drugs available for doctors to prescribe and the amounts in which they prescribe them, despite the fact that there has been no increase in acute or chronic illness, injury, or percentage of persons over the age of 65 years over the same period of time. Drugs have been developed at a remarkable rate during the twentieth century, and in many ways they have contributed to the control or eradication of some diseases. But, in recent years we have also become aware of some negative consequences of frequent drug use (such as side-effects, drug interactions, foetal deformities, or addiction). Most Western countries can now be described as overmedicated with drugs prescribed by physicians who do not have sufficient knowledge about these drugs.[4]

People in the past were not without their drugs, and even thousands of years ago people practised folk medicine using various remedies. Some of these are still with us today, and in some parts of the world, they remain the dominant form of medicine. It is not surprising that herbal remedies persist in Third World countries since at least three-quarters of their population have no access to modern drugs, although there may also be other reasons for adherence to folk medicines. In China, for example, medicinal herbs are used extensively by medical doctors to treat a range of diseases (such as cancer, arthritis, cardiovascular diseases and pneumonia) because they are considered less toxic than modern drugs and in some cases more effective.[5] Some folk medicine was clearly based on superstition, and if it cured at all it did so through placebo[6] effects which, incidentally, may also be said for some of the practises in modern medicine! Some folk remedies however were, and are, certainly biologically successful. These folk cures were developed over years of trial and error experimentation; indeed, many folk medicines have been tested on more human subjects than are modern drugs at their time of release. Only the best of the folk medicinal remedies have survived the test of time, some of them to become the basis for modern pharmacological preparations. Pharmacologists have extracted and tested the active constituents of many of the moulds and herbs used in folk medicine and made them available for prescription in measured doses. For example, the plant *Atropa belladonna*, deadly nightshade, was known to the ancient Hindus and used by physicians for centuries. In 1831 the drug, atropine, was extracted in pure form from this plant, and is still used today.[7] Similarly, reserpine, which is used to treat hypertension and was until quite recently used as a major tranquilliser, was discovered as an extract of an Indian medicinal herb dating back to the ancient Hindus, who used an infusion of it to treat snake-bite, hypertension, insomnia and madness. Many modern drugs are extracts of plants (e.g. the cardiac drug digitalin from foxglove) or micro-organisms (e.g. penicillin from a mould of the genus *Penicillium*). New drugs have been developed from some of these extracts by structural modification and study of their biochemical action. Those who would sneer at all folk remedies should be reminded of this aspect of pharmacology's history, a process of drug discovery which still continues.

Other drugs in modern use have been discovered by careful design based on knowledge of the biological factors which are malfunctioning in a given diseased condition. Consider, for example, diabetes mellitus, a condition in which glucose is incorrectly metabolised. There are two forms of diabetes mellitus, types I and II. Once it was known that the biological symptoms of type I diabetes were caused by insufficient release of the hormone insulin from the pancreas, the drug insulin (extracted from sheep pancreas or prepared by synthesis) could be administered and the symptoms alleviated. Note that administration of insulin does not *cure* the disease: the patient must continue to take insulin throughout his/her life. Only a minority of diabetes mellitus sufferers have the type I form of disease however; some 90 per cent of diabetics suffer from the type II form. This is also known as non-insulin dependent or maturity-onset diabetes. In type II diabetes, insulin release in response to elevated blood sugar is sluggish and treatment is with drugs (such as tolbutamide) which stimulate the pancreatic release of insulin.[8] This does not cure the disease. While the drug therapy manipulates cellular events associated with the disease, it does not provide a unitary explanation of the cause of the disease. Indeed, there are many factors, at different levels of complexity, which correlate with type II diabetes. Diet and physical activity (i.e. life-style) play a role, and these may interact with some genetic constitutions. Increased levels of diabetes occur, for example, in Australian aborigines when they take up a high fat and carbohydrate Western diet, and have less physical activity.[9] These factors are in turn dependent on destruction of their own society and resources and the subsequent low socio-economic status to which they are forced in white Australian society. Although the cure or prevention of diabetes is not so simple as writing a prescription for a drug, in practice treatment for diseases like this, and many more, stops at the point of drug prescription and supply. Many members of the medical profession look no further than the individual, or even the cells and tissues which make up the individual, and believe that by supplying insulin in cases of type I, or stimulating its release in cases of type II, they have focused on the cause of the disease. In reality, they have treated only the symptoms. In other words, the interaction between complex social, economic and possible genetic factors which somehow precipitates the cellular dysfunctions symptomatic of diabetes has been *reduced* (in the doctor's mind) to simple causal explanation at the cellular level. Working on such a simplified and incorrect analysis of the situation he/she mistakenly believes that the disease has been treated adequately by merely providing a drug which suppresses the symptoms.

Virtually every disease can be explained at various levels of complexity: at the societal, the individual, the cellular and the molecular. Each level of explanation can provide causal explanations within that particular level of complexity and must correlate with, or be consistent with, explanations at the other levels. But explanations at one level of complexity cannot be reduced to explanations at a lower level of complexity (see chapter 1). To focus on the simplest level of analysis (the cellular in our example) to provide a causal explanation, without consideration of the other levels, is reductionist thinking.

Causal explanations at the cellular level are insufficient scientific explanations for diseases: they tell only a small part of the story which has serious effects on the forms of treatment that are made available. It is this sort of reductionist thinking, so widely practised by the medical professions of the Western world, that has contributed to the enormous explosion in prescription of drugs. It is not that reductionist thinking in medicine has been without scientific use – it has given us knowledge of at least some of the factors instrumental in disease conditions – but it has focused medical therapy on the treatment of symptoms rather than prevention, or on both jointly. Drug therapy can provide short-term cures, but we need to couple drug use with prevention of the disease that required curing in the first place.

To be fair, Western medicine recognises the need for prevention and has frequently instigated preventive programmes, particularly when a disease is known to be caused, at the biological level, by a specific organism. For example, in some areas of the world where malaria was endemic, such as Italy, malaria sufferers were treated with drugs at the same time that a preventative programme to destroy the mosquitoes carrying the malaria was implemented. However, these disease vectors were destroyed by insecticidal sprays, which themselves have toxic effects on human health, as well as destroying other insects in the ecosystem. Thus, this was a preventative programme which focused on a single factor seen to be causing the disease (the mosquito carrier), against which there was a two-pronged attack of drug treatment of people and destruction of the mosquito. A less reductionist examination of the situation might have led the health authorities to consider agricultural and social practice which were allowing large numbers of mosquitoes to survive in these areas thus increasing people's exposure to them. Such an approach would have led to a several-pronged preventative attack, including methods less dangerous than spraying: placing species of fish which eat mosquitoes into stagnant water; allowing the people to organise a political and economic system which allowed good irrigation and drainage. But programmes like this are not lucrative to large drug and chemical companies!

There is no such thing as a simple cure and prevention of disease even where the immediate biological causes of the cellular events of the disease are known. All diseases have a multiplicity of contributing factors at various levels of explanation, all of which must be considered. The example of malaria was chosen for its simplicity. Treatment becomes much more complex when there is little or no idea of the cellular mechanisms underlying the symptoms of the disease, and even less knowledge about the societal and economic factors that must certainly play a role. The supreme examples of this occur in the so-called 'mental diseases' which are treated by administering psychoactive drugs to individuals. These drugs are major and minor tranquillisers, sedatives, anti-depressants and psychostimulants, which affect the behaviour and psychological functioning of the individual.

Psychoactive drugs

The psychoactive drugs are the most widely prescribed drugs in the Western world; each year their prescription rate increases faster than that for other drugs. A British study of drugs prescribed by 19 GPs over one year in the mid-1970s, for example, showed that psychoactive drugs were prescribed more than any other class of drugs; they made up 20 per cent of the total amount of drugs prescribed.[10] A survey in San Francisco in the late 1960s similarly showed that 20–30 per cent of the sampled population had been prescribed a psychoactive drug in the previous year.[11]

In most Western countries for which we have figures the psychoactive drugs known as minor tranquillisers and sedative-hypnotics make up 80 per cent of the psychoactive drugs prescribed.[12] In fact, one minor tranquilliser, diazepam (Valium), takes the prize for being the most widely prescribed drug in many countries. As prescription of barbiturates has declined, diazepam has more than taken over for treatment of the same symptoms.[13] Indeed, one could well argue that diazepam has filled the place vacated some years before by opium, which was once taken widely and legally by the upper and middle classes for sleeplessness and other minor ailments, and by the working class to make factory life more bearable.[14]

Minor tranquillisers are clearly good profit-makers for the drug industry, and are prescribed in enormous quantities despite the fact that we know very little about how or where they work in the brain. In fact study of the possible mode of action of minor tranquillisers is one of the most challenging areas of research to psychopharmacologists.[15]

Studies of the mode of action of these drugs are, of course, extremely necessary, but they may tell us nothing about the cause of the symptoms for which they are prescribed. Even if we discover the cellular events correlated with the symptoms for which minor tranquillisers are prescribed, we may still find that the drugs do not work directly on these self-same cellular events. In other words, minor tranquillisers cannot be used as tools to allow us to deduce the cellular events causing the symptoms for which they are prescribed.

Minor tranquillisers, according to the medical code of practice, are meant to be prescribed to treat 'neurosis' and 'psychosomatic symptoms'. The medical profession defines neurosis as a 'mental disease' with a collection of symptoms that can be treated by drug therapy with or without concurrent psychotherapy. But neurosis does not have such a neat collection of symptoms that allows us to definitively diagnose either its presence or absence in a given patient. A person is said to be neurotic if they are showing exaggerated responses to given events in the real world; for example, too much and too prolonged grief about the loss of a loved one, or too much worry about an examination. The decision about whether the person's response is exaggerated, however, rests with the doctor prescribing the drug; it relies on that individual doctor's perception and assessment of the patient's behaviour and their life situation. Given that most medical schools provide inadequate training in such areas, and that there is in most cases a class or gender difference between the doctor and patient, this sort of decision is unlikely to be accurately made. The wider the gap between the doctor's life-style and that of the patient, the more likely minor tranquillisers are to be prescribed. Given that most physicians have a high social status, and are males in the 20–55 age bracket, it is perhaps not surprising that minor tranquillisers are prescribed in greater numbers to women, to the elderly and to the lower classes.

Women and psychoactive drugs

Currently, about 80–90 per cent of doctors in Western countries are male; although this should be beginning to change as more women are being trained in medical schools.[16] These predominantly male physicians prescribe twice as many psychoactive drugs to women as they do to men. A survey of more than a thousand middle-aged subjects in West Germany, for example, revealed that 15 per cent of men and 27 per cent of women were taking tranquillisers on a regular basis.[17] Numerous studies of sex differences in prescription of tranquillisers and sedatives have been conducted and consistently the ratio is 2:1 for women to men in affluent Western countries.[18] On the other hand, more men than women are heavy drinkers of alcohol, and fewer men visit physicians to report emotional, financial and life-style difficulties.[19] It seems therefore, that men attempt to solve their emotional and social problems in the social arena amongst peers and by self-medication with alcohol, while women are more likely to seek help from physicians.

However, this can only be a small part of the explanation since Cooperstock, for example,[20] has reported that the number of male versus female visits per year to physicians is 100 to 114 per 100 persons in Ontario; whereas in the same sample twice as many women as men received prescriptions for psychoactive drugs. Other studies have found that with the same number of symptoms reported men are less likely to be prescribed medication than are women.[21] In other words, women's problems are more likely to be medicalised.

In the last century women's problems were considered to stem from their reproductive organs; hence the common diagnosis of 'hysteria', stemming from the uterus.[22] Today, women's problems are often considered to stem from their 'weaker' central nervous systems! The prescription patterns of male physicians are merely a modern-day reflection of the way in which women's feelings and problems are trivialised. The medical profession operates with a medical model which incorporates a variety of culture-bound views of women, and the drug companies have provided them with the tools to do this. The emphasis by drug companies on the female image in advertisements for psychoactive drugs is a powerful reinforcement of these practices.

The wider the gap between the doctor and the patient the greater the probability that a social problem will be 'medicalised' and thus considered to have an underlying biological cause treatable with drugs. If a woman feels unhappy because she has to be a housewife at home alone all day cooking, washing, and so on, it is often easier to prescribe a drug which may help her adjust to this position rather than say, to help her to gain some training and find a job. The decision to make her conform to the role of housewife is also one which usually suits the husband and, at least for him, maintains marital harmony. Prescription of minor tranquillisers has controlled or stabilised the situation, made the husband happy, and provided profit for the drug company. This medicalisation of women's problems is also exemplified by oestrogen therapy given to women during and after menopause (see chapter 3).

Some have argued that women who go out to work suffer more stress, and are therefore more likely to be prescribed tranquillisers or sedatives as a result of 'role conflict' in trying to maintain the mother/wife role while being employed outside the home. The data does not support this view however. In fact, married women living in the 'traditional' female role of housewife take more psychoactive drugs than do married women who work or single women. Other research has shown that married women employed outside the home have lower scores of depression symptoms than do married women who stay at home.[23] These data also demonstrate that explanations based on the assumption that sex differences in biology cause male–female differences in 'psychiatric morbidity' are quite incorrect.

Psychoactive drugs and the elderly

The frequency with which psychoactive drugs are prescribed increases dramatically with age, such that elderly women receive more prescriptions for these drugs than do any other group. Over half the consultations in which prescriptions for psychoactive drugs are written are for patients over 50 years of age.[24] A higher percentage of elderly people visit the physician, but this does not explain why more of them receive psychoactive drugs: studies have shown that whereas 80–90 per cent of physicians expect patients to want prescriptions, only 30–50 per cent of the patients actually do. Most patients would prefer examination, advice and reassurance, but they get a prescription instead.[25] The elderly are the main victims of the modern drug era. Three-quarters of the population over 75 years old receive drugs of some kind, two-thirds receive up to three drugs simultaneously, and as many as one-third receive four to six drugs simultaneously! It is, therefore, little surprise that 10 per cent of the admissions to geriatric wards are due to iatrogenic diseases caused by drugs.[26] Little is known about interactions between various drugs, but certainly some elderly patients receive more than one psychoactive drug simultaneously. This makes pharmocological nonsense, and can be seen as nothing less than a means of social control. Quite apart from the social implications of this kind of treatment, such practices are biologically unsound because elderly people have increased sensitivity to drugs.

Psychoactive drugs and social class:
A relationship between social class and prescription of psychoactive drugs appears to be less consistent (or at least less easily substantiated by most methods of questionnaire assessment) than that between age and prescription rate. Large national surveys in the USA and Australia have found that the highest proportion of long-term use of psychoactive drugs is found amongst subjects with the least education and lowest socio-economic status.[27]

People in the lower classes are less likely to seek professional advice for social and personal problems. However, when they do seek professional help they are more likely to be diagnosed as having a psychiatric disorder, particularly of a 'depressive nature', and more likely to be prescribed long-term use of psychoactive drugs.[28] This is similar to the pattern of treatment for elderly people, and must reflect the social class difference between physician and working-class patients. Working-class people are more likely to be diagnosed as chronically mentally ill and more likely to have their social problems medicalised if they visit a physician for problems other than physical illness.

Psychoactive drugs and the treatment of mental illness

Minor tranquillisers are also prescribed for what are diagnosed as psychosomatic diseases; that is, physical symptoms for which the doctor can find no obvious biological cause either because he/she has failed to look far enough or apply the correct tests, or because there is as yet no known biological cause. One cannot deny that state of mind can correlate with physical symptoms of disease and may sometimes play an extremely important role in them, but it is not possible to divide physical symptoms into those caused by the body and those caused by the mind. In other words, the diagnosis which says that these physical symptoms are imaginary and therefore symptoms of neurosis, must always be incorrect. To attribute the cause of physical symptoms to a biological abnormality in the brain which must be treated by drugs is clearly shifting the reductionist focal point from body to mind while still adhering to the same incorrect analysis of cause.

There is no real definition of a 'mental disease'. A disease is something which can be defined by a characteristic collection of symptoms, some or all of which must occur in the patient. So-called 'mental diseases' have no such clear-cut symptoms, and the loose collection of symptoms used to diagnose them can clearly lead to many people being mistakenly diagnosed. Thus, given a growing dependence on drug therapy to treat 'mental illness', it is hardly surprising that prescription of minor tranquillisers continues to rise. Any individual who feels unhappy in their acquired social role and goes to a doctor for help is an open target for minor tranquillisers or sedative therapy. The individuals in society most likely to feel this need are those in oppressed groups, the men and women in lower classes, and women and elderly people in all social classes.

The term 'mental disease' implies that the cause is inherent in the individual diagnosed as mentally ill, and indeed caused by some abnormality in the biological functioning of that person's brain. It underscores the belief that, even though we do not yet know of a cellular cause for neurosis, one day medical science will provide us with such a cause and we will then know exactly how to treat the disease by manipulating cellular events. But even though we may discover cellular processes which correlate with certain mental conditions, there are also many other levels of explanation for any given mental state. The sole cause is not rooted in the cellular events. Treatment aimed solely at manipulating biology is therefore incorrect and dangerous.

This is not to argue against all use of psychoactive drugs. For a limited number of people who have been allowed to progress to such an extreme degree of disturbance that they can no longer communicate with other people or may be tortured by very bizarre hallucinations, temporary drug therapy may assist them to return to a mental state which can allow them to start resolving some of their problems through some sort of psychotherapy[29] coupled with broader social and political changes. The answer for these people, who are usually labelled as 'psychotic', is not to be locked up in a psychiatric ward or to receive treatment with major tranquillisers for long periods of time (often amounting to several years or even the rest of their lives). Locking people up in psychiatric wards is merely a form of imprisonment which serves to protect society from the so-called 'mentally ill'. Use of psychoactive drugs over the last three decades has meant that fewer people have been forced into long-term institutionalisation, but long-term drug therapy can also have the effect of locking a person away from playing a full part in society.

First of all, major tranquillisers do not cure. They may block nervous transmission along particular brain pathways (the dopaminergic pathways)[30] but this does not mean that psychosis is caused by abnormally high levels of dopaminergic activity in the brain, or for that matter by any other aspect of the biochemical action of major tranquillisers.

Second, major tranquillisers have many side-effects. Pharmacologists are the first to advise that major tranquillisers should be prescribed for only brief periods of weeks and not months or years, and at the lowest doses possible in order to avoid side effects – of which there are more than a dozen that commonly occur. These include postural hypotension (decreased blood pressure immediately after standing up), amenorrhoera (ceasing to menstruate), liver dysfunction, photosensitivity and, in particular, motor dysfunctions.[31] Parkinsonian symptoms (motor dysfunctions such as tremor, 'pill rolling' motion of the fingers, restlessness of the legs) frequently occur soon after major tranquilliser treatment commences, and are more likely to occur with higher doses. Patients showing these symptoms are usually treated by prescribing an anti-cholinergic drug to be taken along with the major tranquilliser. Some psychiatrists routinely prescribe an anti-cholinergic with major tranquillisers particularly if they prescribe high, poorly adjusted doses of the latter.

With prolonged use of major tranquillisers, especially at higher doses, an even more debilitating form of motor dysfunction can occur. This is tardive dyskinesia, which includes repetitive spasm of oral and facial muscles, and involuntary motor movements. It occurs in about 25 per cent of patients on long-term therapy.[32] There is no adequate treatment for this condition, and once precipitated by the drug therapy it usually persists even after the patient has stopped taking the drug. Anti-cholinergic drugs exacerbate tardive dyskinesia. Tardive dyskinesic symptoms can be temporarily suppressed by increasing the dose of major tranquilliser, but this increases the probability of Parkinsonian symptoms reappearing. Clearly, prolonged use of major tranquillisers to treat psychosis sets up a vicious circle which cannot be justified on pharmacological grounds. However, major tranquillisers are frequently prescribed without any, or with insufficient, psychotherapy thus necessitating their prolonged use by the individuals who are being made to conform to a society in which they no longer feel the urge to work towards social change.

Elderly people have a much greater probability of developing tardive dyskinesia when treated with major tranquillisers. It occurs in up to 40 per cent of elderly patients, and in women more often than men.[33] Yet it is to the elderly that major tranquillisers are most frequently prescribed, even after casual observation by the physician. Furthermore, when psychoactive drugs are prescribed for elderly people, little or no consideration is given to possible interaction with other drugs which they may be taking (e.g. cardiac drugs). In fact, not very much is known about how such drugs interact.

The contributions made by elderly people to society are largely ignored in societies with Western capitalist economies. It is not surprising then that they have problems which result from their devalued social position. Drug therapy acts to dull their feelings of discontent, and removes their desire to fight for social change. Older people are also considered to be physically deteriorating. Therefore social problems and the unhappiness which they feel are more likely to be tied down to deteriorating biological function of the brain. 'Senility' is considered to be a diagnosable 'disease' stemming from cellular malfunction. Elderly people are frequently diagnosed as having 'senile dementia' or 'organic brain dysfunction' simply on the basis of disorders of cognitive function.[34] Sometimes brain lesions are detected in association with this diagnosis, but many individuals are diagnosed as having 'organic brain dysfunction' when no physical abnormalities have been detected. The result is that elderly patients accept medicalised explanations for their behaviour as part of that which they see as their inevitable decline. The medical label has put them on the scrap heap, the drug treatment that they usually get thereafter makes certain that they never get off it.

Although there are limited cases where psychoactive drugs can be usefully, if briefly, applied, there is far too much incorrect diagnosis and a vast amount of overprescription of these drugs.

We have discussed above the medicalisation of social and behaviour problems which contribute to overuse of psychoactive drugs. The role of the individual doctor has been highlighted in overprescription of psychoactive drugs, but the practice of individual doctors should not be seen as the sole cause. Indeed, individual doctors are merely instruments in a process which begins with multinational drug companies seeking profit. Below we will follow the process of how a given type of psychoactive drug is discovered and marketed.

Launching new drugs

In the 1930s a chemist called Sternbach considered that the chemicals known as 4, 5–benzo–[hept]– 1, 2, 6–oxdiazepines might be useful as a starting point for developing a new series of dyes. This hope did not materialise, but Sternbach persisted with these compounds until, many years later when employed by Hoffman La Roche Laboratories in Nutley, New Jersey, he developed a series of benzodiazepine chemicals which could be used as minor tranquillisers. Chlordiazepoxide was the first to be released on the market as a minor tranquilliser, in the 1960s. Very soon, after only a few years of clinical trials, it became the most highly prescribed drug for treatment for 'neurosis and anxiety'. During the last ten years it has been largely replaced by another derivative, diazepam (Valium), which is more potent. The drug companies with patents for these drugs captured the markets and have consequently made enormous profits.

The enormous size of the market for the benzodiazepines was dependent on many factors. First, these drugs had to be promoted, and the drug companies put vast quantities of money into advertising them. This promotion of the benzodiazepines obviously did not occur in a void. It occurred in societies where individuals had already accepted the belief that if they were unhappy with their lives then it was *their* fault as individuals, and *they* were the ones who had to change. It also occurred at a time when the taking of such drugs prescribed by the medical profession was, as it still is, seen as a completely socially acceptable thing to do. The 1960s was a time when people in the affluent West were experimenting with a range of social drugs, some self-administered, others prescribed by the doctor.

The people of the Western world were thus prepared for new drugs, and advertising by the drug companies fell on fertile ground both inside the surgery and in wider society. Teaching in medical schools promoted these drugs, as it continues to do today, and medical students were (and are) so firmly taught reductionist theories for the 'cause' of 'mental disease' that they came to accept them as truisms.

Advertising of psychoactive drugs is grossly reductionist. It medicalises complex human behaviour in such an overconfident fashion that it seems to have all the answers. For example, 'Sweet marital bliss has turned sour, and everything seems beyond her capabilities. Sinequan could help her cope'; or the caption 'Serepax keeps working patients active' under a photograph of a housewife vacuuming[35]. These examples serve to reiterate a point made earlier – advertisements for drugs have women as their target, and especially women in the traditional role of married housewife. Other advertisements use the common technique of featuring the nude female as an eye-catcher, and this again links women with the prescription of these drugs. Such advertisements are written by male-dominated companies for male doctors, both of whom stand to gain from sale of the product. A study of 500 drug advertisements selected from over seven years' publications of the two most widely distributed medical journals in Australia found that advertisements for psychoactive drugs used pictures and they mostly portrayed female patients. Male and female pictures appeared with equal frequency in advertisements for non-psychoactive drugs[36]. Most of these glossy advertisements are severely lacking in pharmacological content.

A similar survey of advertisements for psychoactive drugs appearing in medical journals in the USA, found that male patients in advertisements were more frequently associated with rational statements about drug dose, mode of action, safety, and so on; whereas, female patients were associated with irrational statements appealing to the physician's ego, humour, sexuality or empathy with a suffering, weak-minded patient.[37] Thus advertising by drug companies reproduces and reinforces the ideology which preserves the social order in which they flourish.

In addition, the drug industry, partly through advertisement, is ingenious at inventing diseases and cures for them.

Hemminiki[38] has suggested that doctors create illness in order to treat it with psychoactive drugs, and they prescribe those drugs in order to justify psychiatric diagnoses. It is arguable that the source of this creation is closer to the drug companies with the health services and physicians being their willing instruments. Psychiatric diagnosis is, in fact, frequently tailored to suit the drugs, because diagnosis is often made after trying a number of classes of psychoactive drugs and seeing which ones best suppress the symptoms being monitored by the physician.[39]

As mentioned earlier, the social classes being trained to be doctors and the social class attained by doctors, establishes a doctor – patient gap which serves to increase the amount of psychoactive drug prescription. This is a reflection of the education system. Education is not freely available to all in capitalist countries, and very few patients prescribed psychoactive drugs know how they are meant to work or, perhaps more to the point, how little is known about how they act. Also, few patients know that the drugs simply suppress symptoms and do not cure. Even fewer are aware of the dangers of addiction, physical or psychological. There is evidence that minor tranquillisers can be both physically and psychologically addictive[40]. Addiction also increases the sales of these drugs. Finally, we live at a time when people in Western cultures hold a utopian view of health[41]. Unlike generations before us, we expect to be at about the peak of health most of our lives, and we expect to get treated, to take drugs, to keep us in this state. Many diseases like 'flu or colds will get better by themselves, but most people seek medical advice and demand drug therapy for them. Overprescription of antibiotics at incorrect doses has favoured the production of mutant bacteria which are resistant to nearly all forms of antibiotic and present a very serious problem in hospitals.[42]

Similarly, people expect to remain in some utopian state of mental happiness or optimal mental functioning by taking medication to achieve this.[43] If they fail to get an expected number of hours sleep at night, they take hypnotic tablets. This is particularly prevalent in elderly people who expect, and are told to expect, to sleep as long as they did when younger even though they may be physically less active than they used to be. Taking a sleeping pill at night can make some people drowsy in the mornings; so they take another pill to pep themselves up again, and so it goes on. Utopian belief in constant happiness of the bland accepting kind has been sold to us by a society (or rather by those who hold the power in society) which wants to remain as it is, uncontested with everyone fitting into their designated social role and happily accepting their designated economic status.

It is true that many individuals feel at odds with Western capitalist society, and increasingly so as unemployment grows and social welfare diminishes; but the problem is not with them or their biology. These problems will not go away by treating individuals either with psychoactive drugs or any form of psychotherapy aimed at the individual. It is in the interests of the minority who benefit most from the present social organisation to suppress opposition, and one of the best ways to do that is by issuing the vast quantities of psychoactive drugs which we see distributed in Western society today. What a convenient situation this is; the tools used to maintain the present inequitable system also provide enormous financial profits for those companies at the top of the hierarchy.

5. Western scientific medicine: a philosophical and political prognosis*

Len Doyal and Lesley Doyal

Introduction

At a time when the defence of health services has become a major focus of political struggle, modern medical science has itself been vigorously criticised. Some argue that through emphasising chemical and surgical treatment of individuals, attention is distracted from the fact that much illness is environmental in origin and therefore potentially preventable. Others point out that not only is modern medicine less effective than its proponents would have us believe, but that it can itself constitute a threat to health. Still others claim that medicine tends to view people as objects to be manipulated. They point out that much medical thought and practice splits the physical and the psychological – the body and the mind – thus importing an unnecessary dualism into our conception of the 'self'. As a result, medical treatment is said to be less effective and more alienating than might otherwise be the case. This chapter attempts to render these criticisms more philosophically and historically understandable, and to assess their underlying political and social significance.

The scientific revolution and the mechanisation of nature

It is necessary to begin by considering the revolution in human understanding that shook the Western world during the seventeenth and eighteenth centuries. Prior to this

* The authors would like to thank Mary Ann Elston, Mike Joffe, and John Farquahar and Grenville Wall for their help in preparing this article.

time, the open universe we now take for granted was seen as a closed world with the Earth at its literal and figurative centre. The vast majority of medieval Europeans made sense of their everyday lives in what were essentially supernaturalistic terms. However, for the small minority of literate clergy and scholars, religion combined with classical Greek thought to provide a powerful cosmology for explaining nature and society. Within this broadly Aristotelian framework, it was assumed that pure forms of earth, water, air and fire were the elemental building blocks of reality. Each of these elements had its own qualities of dryness, wetness, coldness and warmth which were combined in various proportions to make up the natural world.

Within this world view, the main concept used to explain particular events was that of purpose.[1] Thus questions like, why do girls grow into women, acorns turn into oaks, heavy bodies fall to Earth, fires rise from it, and slaves serve masters? were all answered by reference to some actual or potential state of affairs which was defined as 'natural' and which it was the *purpose* of all parts of nature either to maintain or to achieve. Solid bodies were said to fall when dropped and the flames of a fire to rise because their respective 'natural places' were below and above the Earth. Girls were said to grow into women and acorns into oaks because it was their 'natural potential' so to do, just as it was the 'purpose' of slaves not to be free but to serve someone naturally predisposed to be a master. Similarly, health was seen as the 'natural state' of a human organism. Illness was believed to be the outcome of a disruption in the normal balance of the person concerned, so that treatment should consist largely of helping the body's own purposeful attempts to restore itself to natural functioning.

We are all familiar with this Greek/medieval explanatory approach. We use it, for instance, when we say that an action requires no further explanation because it is 'just human nature' or that organisms possess particular internal organs (e.g. lungs) *because of* their biological function or purpose (in this case respiration). But the problem with understanding nature or society in terms of purpose alone is that it provides us with no means by which to *control* them –

a task that was being taken increasingly seriously as the scientific revolution began in the early seventeenth century. Knowing the purpose a new technology should serve does *not* tell us how to achieve it. Moreover, if we do eventually manage to innovate successfully, a purposive explanation will give no indication of *how* we were able to do so.

In response to such problems, a new mode of understanding was gradually developed which later became the cornerstone of modern natural science. This entitled the observation and description of physical regularities, and formulation of 'laws', which could then be applied to either natural or artificially created situations in order to predict their outcome, and ultimately, to control them. It was necessary for instance to establish the laws of terrestrial and celestial motion, and those of the behaviour of chemicals and metals in order to improve the important skills of navigation and weaponry. This emphasis on physical laws led eventually to the belief that nature could be likened to a machine consisting of interlocking parts with regular, reoccurring relationships. Thus the 'mechanisation of nature' led away from the purposeful or 'teleological' explanations of the past towards the belief that all physical phenomena must be understood with reference to their 'causes' – those general circumstances, abstracted by laws which seem to necessitate their specific occurrence.[2]

In studies of motion, light, electricity, magnetism and various aspects of chemistry, this 'deterministic' explanatory approach soon led to discoveries that were as dramatic as they were numerous. Indeed, by the end of the nineteenth century, many European intellectuals believed that most of the basic truths about nature had already been discovered.[3] This optimism was also applied to the study of human physiology. As early as 1628, William Harvey had demonstrated that the movement of blood round the body could be understood in hydraulic terms with the heart acting as a pump and the veins and arteries as connected pipes and tubes constituting a circulatory system.[4] Similar analogies were then formulated to enable the successful understanding of muscular, reproductive and other bodily systems. Practical research in anatomy produced such convincing results that by 1800 the mechanistic conception of the human body had become a commonplace, and most experimental work continued to be conducted within that overall framework. However, the application of this machine-model to conscious human beings was very different from its use in explaining the behaviour of inanimate objects or even animals, so that from the earliest times, the implications of these new developments were hotly debated.

Of all seventeenth century writers, it was the French philosopher René Descartes who most clearly recognised the philosophical implications of this new anatomical science. Descartes agreed that the physical body should be understood as a machine but argued that there were other parts of the person that could not be explained in this way. These he denoted by the expression 'mind' which was to include various aspects of human consciousness from the passive experience of sensation to active and rational deliberation. Descartes' main argument was that in almost every respect, the mind and the body possessed opposite characteristics. Minds could not be seen or touched whereas bodies were only too visible and tangible. Moreover, minds could make choices of a sort entirely inconsistent with the tenets of determinism, involving as they did, clear evidence of the 'free will' necessary for the attribution of moral responsibility.[5]

Consider for a moment, how we explain and evaluate a human *action*. We see it as a decision by a mind *to do* something using the body. Ostensibly it is the reasons – the aims towards which the action is directed and the strategy stipulated for achieving them – which makes what is done intelligible and constitutes the basis for ethical evaluation. Machines, on the other hand, are explained in terms of the causal interrelation of their parts or the active manipulation of human agents. It is in this sense that we might readily hold a *person* responsible for a crime but would not do the same for a robot![6] In short, Descartes divided the person into a material body in which things happen (e.g. the operation of the circulatory system) and a non-material mind which does things (e.g. makes the decision to perform or to undergo surgery). He further argued that this dualism meant that the mind and the body are independent sorts of things, the mental and the physical having no causal relationship to each other of the sort which physical objects have.

There were many problems with Cartesian dualism, not the least of which being the question of how something non-material could influence the behaviour of something material. Of course he accepted that the mind and the body *do* interact somehow but the *nature* of that interaction remained a mystery. An action as simple as raising one's hand seems inexplicable on these grounds – the hand having no *contact* with the reason for raising it – but obviously it happens. Moreover, despite the differences between the 'mentalistic' language of giving reasons for actions and the 'physicalistic' language of giving causes for events, both seem necessary for a complete understanding of our relationships with each other and with the physical world. Indeed it is through the use of both – through descriptions of specific bodies associated with specific consciousness – that we actually distinguish one person from another.

As a result some form of dualism has set the pattern for most debates about what it means to be a person (as opposed to just an animal) since the time of Descartes.[7] But what are the implications of this for medicine – where ostensibly it is more difficult to separate the mental from the physical since we depend so much on the actor/patient for information? For were the mind and body really so separate, the reliability and diagnostic relevance of such information would be called into question. This is precisely what began to happen in the eighteenth century.

Dualism and the mechanisation of medicine

During the seventeenth and eighteenth centuries, medicine did not yet enjoy the remarkable success of the physical sciences. This might seem surprising in the light of early progess made in anatomy, but it is significant that during this time the study of anatomy and physiology had little connection with efforts to understand and treat human illness. Certain diseases were loosely correlated with the malfunction of particular organs, but early anatomists made no systematic attempt to demonstrate these correlations empirically – to differentiate, for example, between a diseased liver and a healthy one.

There were various reasons for this. One was undoubtedly the great divide between the university-trained physicians and the much lower status barber/surgeons. But another important element was the nature of the relationship between doctor and patient during this period. In general, doctors were of a *lower* social class than their patients and were dependent on patients' patronage to earn a living. As a result, they were reluctant to undertake the intensive examinations and experimental therapies that empirical investigations would have necessitated. Hence it was in their interest to retain earlier theories of health and illness which emphasised the individuality of their patients while at the same time making the task of empirical research even more difficult.

Borrowing from the Greeks, seventeenth and eighteenth century doctors still explained health purposefully in terms of the 'natural' balance of elemental body components analogous to earth, water, air and fire. Although the precise formulation varied in different schools of academic medicine, those basic elements were the four so-called 'humours' – blood, yellow bile, black bile and phlegm. These humours were thought to be fluid-like substances found throughout the body and each was associated with overall states of being rather than with the functioning of specific organs. Phlegm, for instance, corresponded with the element water and was, therefore, cold and wet and associated with stolidity of temperament. Yellow bile on the other hand corresponded to the element fire, was hot and dry and associated with irritability. If health was to be maintained, the humours had to be kept in a balance that was natural for the individual concerned. When this balance was disturbed, ill-health was the inevitable result and the body would attempt to restore health through ridding itself of the excess of the particular humour responsible.[8]

When we look back at past medical practices such as the leaching of blood, our response may be to laugh at their naivety. But it is important to stress that there was nothing inherently silly about these ideas. Just like modern scientists, Greek and medieval physicians were attempting – within the limits of their technology – to correlate external symptoms with what they thought was going on inside the body. Thus, it is easy to understand why in observing a cold, say, or some other kind of infection, they saw no reason to doubt the humoural theory, since the body does indeed appear to expel foreign matter of the required viscosity and colour, and to cease to do so when health returns!

According to the humoural theory, knowledge of internal bodily states was mostly to be acquired from the patient's own description of her/his symptoms. These had to be detailed and include information about a wide variety of mental and physical experiences. Unless the character of the illness was very obvious (as in the case of smallpox, for instance) the diagnosis was made primarily from these subjective symptoms reported by the patient. Yet great care had to be taken in generalising about the relationship between symptoms and illness in *different* people. For example, while in one person a fever might be the result of an excess of yellow bile, in another it might be caused by a deficiency of phlegm. The correct diagnosis would depend to a great extent on the temperament of the particular patient which in turn depended on the humoural balance which s/he 'naturally' possessed. Humoural theory was, in short, extremely individualistic in character, and was ideally fitted to the sort of medical practice that emphasised the social relationship between doctor and patient. It was essentially 'patient-oriented', stressing the uniqueness of each case and the necessity for personalised treatment.[9]

Not surprisingly, humoural treatment consisted mainly of herbal or chemical purgatives and/or bleeding. Thus, the therapeutic ideal was to aid the body in ridding itself of whatever humoural excess was producing the illness. Other forms of treatment included prescriptions for rest, for dietary changes and for non-purgative herbal medicines employed for their soothing or healing properties. However, many of the purgative treatments in particular were very vigorous, and probably dangerous. Indeed, the lack of predictability and the hazards associated with such treatments often gave practitioners a dubious reputation, and by the middle of the eighteenth century it had become as popular in intellectual circles to mock physicians as it is today to make fun of the failings of economists![10]

One of the main difficulties with this pre-scientific medicine was that the results of its treatments were not reliable. This was related to the fact that few of its hypotheses could be empirically tested. Such tests were inevitably problematic so long as illness was understood primarily through the idiosyncratic reports of individual patients. Even with common ailments such as 'consumption', different patients would not experience the same symptoms, especially as far as the psychological dimension of the illness was concerned. There was, therefore, considerable doubt about how – and indeed whether – 'consumption' could ever be clearly defined as a clinical entity. This meant that even if treatment *appeared* to work in one case, it was never clear whether it would also be effective in helping others apparently suffering from the 'same' diseases.

Only an accurate classification of diseases in terms of symptoms that were universally applicable to all sufferers in more or less the same manner could change this situation. But for such a classification scheme to be evolved, conceptions of disease had to become much more specific and subject to agreed criteria of identity that were not totally dependent on the conscious perceptions of the sufferer.[11] The process by which this was eventually achieved was a long and complicated one. It can be divided into two main stages, the first involving diagnosis and the second treatment.

In the early 1760s, an Italian named Morgagni (amongst others) was advocating the correlation of reported symptoms with observable anatomical lesions (abnormal features of particular organs). He was one of the first to show how this could be done through dissection after death, thus beginning the documentation of the physiological effects of particular illnesses on different parts of the body. Yet this still left the problem of the idiosyncratic identification of symptoms, since a generally accepted classification of the symptoms associated with particular diseases was just as important as the anatomical classification of organic pathology.

With this in mind, new techniques of observation were developed which could identify symptoms *without* depending on the subjective perceptions of the patient. By the mid-nineteenth century the stethoscope, the ophthalmoscope and the laryngoscope had all been introduced to provide diagnostic information about different parts of the body. These new observational methods were given a significant boost in 1895 with the discovery of X-rays by Roentgen, making it possible for the first time to develop a classification of the pathology of organs in the *living* body. Microscope technology also continued to progress while by the end of the nineteenth century, machines had been invented for measuring pulse rates, blood pressure and heart beat, and chemical procedures had been designed to identify abnormalities in blood and urine. In this way, the diagnostic importance of the patient's own perception of her/his illness was gradually minimised. Increasingly, diagnostic information was (and is) acquired from machines which monitor particular physiological processes and generate the appropriate empirical data – often without the understanding or sometimes even the awareness of the patient concerned.[12]

This new technology further encouraged the belief that diseases should not be seen as systemic 'imbalances' as they had been in the past, but as highly specific disorders affecting the micro-structure – the tissues – of particular organs. This emphasis on 'specific aetiology' in turn supported the belief that wherever possible treatment should consist of highly specialised chemical or surgical intervention. At the same time, the isolation of the tubercle bacillus by Koch in the 1880s, and the elaboration of the germ theory of disease by Pasteur, reinforced the belief that much illness was caused by specific biological pathogens. This made it plausible to search for effective treatment in isolation from what the patient actually thought or felt – either by chemically destroying the organisms or immunising people against their effects. This search for chemical cures of particular disorders has been characterised as the quest for 'magic bullets'. It was in this context that the science of bacteriology evolved, giving the search for effective therapies a firm experimental grounding. By the beginning of the twentieth century a long list of diseases had been linked with specific micro-organisms identified in the laboratory (tuberculosis, diptheria and syphilis, for example). These developments culminated eventually in a number of effective vaccines and in the introduction of antibiotics in the late 1940s.[13]

These developments in medicine in the nineteenth and twentieth centuries provided valuable therapies – vaccination, antisepsis and anaesthesia making surgery safe, drugs to relieve pain, and antibiotics to treat infections, were probably the most significant. It is important that these genuine advances are not forgotten – a point to which we shall return later. However, at the same time, it is also clear that in the process of achieving these successes, the nature of medical knowledge and medical practice was transformed in ways that were not necessarily conducive to the therapeutic endeavour. It is in this context that we can understand, for instance, why modern medicine is often said to treat the disease and not the patient. When a sick person enters hospital, s/he is interviewed mainly to obtain specific information for a more complete medical diagnosis. From the doctor's point of view, what the patient feels about her/his predicament is often seen as only minimally relevant to the task at hand: the effecting of a cure. Indeed the entire success of the enterprise is thought to depend on the patient's maximal 'compliance' or *passive co-operation* – speaking only when spoken to, answering questions as accurately and unemotionally as possible and, above all, doing exactly what s/he is told.[14] The apparent acceptance by both patients and doctors of such impersonal and disease-oriented patterns of diagnosis and treatment is perhaps surprising, and can only be understood against the background of wider social and economic developments.

The division of labour characteristic of the early industrial revolution required a large and disciplined workforce concentrated in the towns. This led initially to widespread poverty and ill-health and one response to these problems was the building of large public hospitals. Whatever the intentions of their original founders, these institutions had the side effect of providing an unending supply of impoverished and largely compliant patients. These 'undeserving sick' had little option but to allow physicians, surgeons and other researchers to do what they liked in the experimental assessment of their new medical technologies, sometimes with frightening and even fatal consequences. These people were the guinea-pigs on which their social superiors (the medical profession) were able to use the new techniques of observation and measurement and to expand their own knowledge and skills.[15]

Thus the development of industrial capitalism provided the social and economic setting within which modern medical science was able to develop. At the same time, however, medicine also aided the consolidation and expansion of this same historical process. This was not so much because of its practical success in helping to maintain a healthy workforce, for as we shall see, its contribution in this respect has been much exaggerated. Rather, its primary role was an ideological one in that it led people to perceive many of the unpleasant aspects of their lives – and deaths – as things over which they could not have control. In the first half of the nineteenth century, the importance of environmental influences on health was widely recognised (if not theoretically understood) and the public health acts of 1848 and 1872 reflected these concerns. Yet by the end of the century the emphasis had been firmly shifted towards curative rather than preventative medicine. The belief that diseases were caused by micro-organisms that attacked people at random, implied that wider social, economic and environmental conditions were not to blame. Individuals who contracted an illness were therefore seen as unlucky, or even as being themselves at fault because of their 'insanitary habits'. In either case, curative medicine rather than social change was said to be the answer, and the 'state of the public health' was placed firmly in the hands of medical 'experts'.[16] This belief in the accidental character of illness still continues today, with more medical intervention advocated as the solution to the so-called 'diseases of affluence' that increasingly plague advanced industrial societies.

Western scientific medicine: an inappropriate technology?

We have described the historical background to what is usually referred to as Western scientific medicine. The 'biomedical model' on which it is based can be summarised in terms of five basic assumptions:

(a) The body is seen as analogous to a machine whose component parts are interrelated and whose functioning is explicable with reference to various disciplines within the natural sciences.

(b) Health is that state of the body where the various component parts are more or less stable in that they are all doing what is necessary for the organism to work successfully. What it means 'to work successfully' is usually defined in terms of fitness to perform one's appropriate social duties.

(c) Illness is therefore defined as a malfunction of one or more of the body's components entailing a set of measurable symptoms, signs and biochemical or physical abnormalities.

(d) Disease is believed to result either from 'degenerative' processes within the body, from biological chemical or physical pathogens invading the body from outside and doing damage to particular organs, or from a failure in one of the body's own regulatory mechanisms (itself unexplained). The damage then spreads either because other organs are similarly invaded or because they are biologically dependent on the first.

(e) Medical treatment consists primarily of attempts to restore the body's normal functioning either by halting 'degenerative' processes, by boosting regulatory mechanisms, or by destroying the invader. In this sense, cure is often conceived as a form of warfare which employs chemical or surgical weapons that have shown themselves to be effective in similar instances in the past.

Medicine based on the biomedical model has had many

notable successes. Indeed, it is hard for us today to appreciate the terror that used to be caused by diseases that are now easily cured and the dread associated with the barbarity and dangers of treatment.[17] Today, we take it for granted that if we contract one of a wide range of illnesses, a relatively safe and effective cure will be available. And, to some extent, we are right to believe it, as anyone who has had an inflamed appendix successfully removed will quickly testify.

However, this does not mean, as is often claimed, that modern medicine was responsible for the dramatic improvement that has taken place in the health of most Western populations over the past hundred years. In this respect at least, its success has been widely exaggerated. Nor does it mean that Western medicine in its current form provides the best possible means for improving or maintaining health. In expanding these points, we will explore further both the ideological character and the practical deficiencies of the biomedical model.

Medicine and public health

The major fatal diseases of nineteenth-century Britain were infectious diseases such as tuberculosis, cholera, diphtheria, measles and scarlet fever. Tuberculosis in particular was very widespread and viewed with much the same dread that cancer is today.[18] Indeed, adjectives like 'tubercular' and 'consumptive' were employed as metaphors in much the same way that 'cancerous' or 'malignant' are often used today. If the orthodox picture of the march of medical progress were correct, we would expect the fall in the death rate from these diseases to have come about as a direct result of medical advances. However, in practice, they were all eliminated as major causes of death *before* the development of successful therapies such as antibiotics. This was because the main factors responsible for the spread and the severity of these diseases were environmental. Contaminated water, poor sanitation, crowded housing without proper heating or ventilation, and inadequate levels of nutrition, all combined to create ideal conditions for the communication of these diseases. Not surprisingly, therefore, it was public health legislation along with a real rise in the standard of living that were primarily responsible for their containment. The incidence of tuberculosis, for instance, had been greatly reduced in the UK long before the advent of modern anti-tubercular drugs in the 1940s.[19] This is not of course to deny the enormous importance of these drugs for those individuals who continue to contract the disease (there are currently estimated to be some fifty million sufferers, mostly in the Third World). However, it does raise serious questions about how such diseases can most effectively be tackled.

The picture is no brighter when we turn to the major fatal diseases of the twentieth century – cancer, heart disease and strokes in particular. Here, cures have not been found, despite the expenditure of vast sums, and a degree of technological sophistication that would have been marvelled at only fifty years ago. Nor, of course, have preventative techniques been developed, comparable with vaccination for smallpox or polio. Cancer is probably the most dreaded of contemporary diseases and surgery, radiotherapy and chemotherapy are all used in its 'treatment'. However, the survival rate for most forms of cancer is hardly greater than it was thirty years ago, despite the billions spent on research in the intervening years. In the case of lung cancer for instance, the average time of survival after diagnosis is only about six months. But as with the infectious diseases of the nineteenth century, it is now widely accepted that successful strategies for the prevention of many cancers *could* be devised. This would involve the creation of an effective system for the identification and regulation of the carcinogens (or cancer-causing substances) that are said to be implicated in the production of about 80–90 per cent of all cases of the disease. Thus, a considerable impact could be made on the future cancer burden, *without* the use of modern medical techniques, and similar strategies could probably be adopted to reduce the incidence of heart disease.[20]

So modern medicine has failed to develop effective treatments for many of the diseases that currently ail us. This failure is probably illustrated most dramatically by the fact that the introduction of curative medicine, in the form of the National Health Service, has not reduced class inequalities in health and illness in the UK. Despite the fact that the NHS led to a much greater equalisation of access to medical care for different social classes, this has not been reflected in relative patterns of morbidity and mortality. A child born in the UK in 1982 whose parents were unskilled workers was about twice as likely to die in its first year as a child born to professional parents. Indeed, working-class people *of all ages* are at a greater risk of dying than their more affluent contemporaries, and suffer a greater incidence of both acute and chronic sickness.[21] The significance of these social class differences is, of course, that while medicine undoubtedly has an important part to play in *caring* for those individuals who become sick, it does not have as great an effect as is often imagined on the general state of the nation's health – particularly by comparison with wider environmental factors such as living and working conditions.

The biomedical model and the individual

When we turn from patterns of disease in whole populations, to look at the experience of individuals, the explanatory power of the biomedical model is somewhat paradoxical. On the one hand, medical techniques can successfully be used in the diagnosis and sometimes the cure of many diseases, thus relieving much individual suffering. Yet in another sense, the biomedical model has been a dramatic failure because it cannot explain why some people do *not* become ill when, all things else being equal, they *should*. Were the doctrine of specific aetiology as powerful as is often suggested, then presumably everyone – or at least the majority of people – in the immediate vicinity of particular organic or inorganic pathogens should develop the appropriate illness. Yet in almost every case they do not. For instance, some miners get pneumoconiosis while others do not, and some people get influenza when 'there's a lot of it about' while the majority do not. Obviously, *some* of these variations are related to easily identifiable differences between individuals and their environments – an asbestos worker who smokes, for instance, is more likely to develop asbestosis than a colleague who does not.

However, differences of this kind cannot provide anything

like a total explanation of this differential response to known pathogens, and there is a growing recognition that the susceptibility of the potential 'host' – the person exposed to the pathogen – plays a significant part.[22] Furthermore there is now increasing evidence to suggest that these variations in susceptibility are not due simply to straightforward immunological differences of a genetic kind but are directly related to many other aspects of the host's life – in particular to the amount of stress s/he has experienced, both at the time of exposure and in the past.[23] It has been argued, for instance, that many diseases show marked variations in accordance with the cyclical fluctuations in the business cycle, and that hypertension and coronary heart disease are more likely to occur in people under stress in low-paid jobs, or among those who have recently become unemployed.[24]

Hence the combination of physiological, psychological, sociological and emotional variables involved in any one individual's reaction to a potentially illness-producing situation is likely to be extremely complex.[25] It is obviously important that we understand how these processes work since this would provide important insights into the very nature of health itself. Yet because of its emphasis on biological determinism and its lack of concern with the impact of the social enviroment on individual consciousness (and lack of consciousness) the biomedical model has little to offer in this respect.

Unhealthy medicine

As well as these therapeutic and explanatory failures, Western scientific medicine has also been accused of being dangerous itself – of producing what are known as 'iatrogenic' (literally, doctor-caused) diseases. Direct iatrogenesis occurs when an illness or disability results from the unhealthiness of the hospital environment (e.g. staphylococcus infections resistant to safe antibiotics), through mistakes in treatment (e.g. poor surgery or anaesthesia), as a side effect of drug therapy (e.g. the thalidomide débâcle), or of diagnostic technology (e.g. possible radiation damage associated with some types of cancer screening). The risks of such procedures must, of course, always be weighed against the severity of the illness and the benefits that the treatment *might* confer. Unfortunately, however, it is impossible for most people to make a reasonably informed choice about such risks, either because the circumstances are unpredictable, or the issues themselves are not properly explained or understood.[26]

The second form of iatrogenesis laid at the door of contemporary medicine is indirect, and paradoxically stems from some of medicine's real successes. It occurs when people depend on doctors or on over-the-counter drugs to suppress their symptoms or to normalise their bodily functions, rather than controlling whatever is creating the 'disregulation' in the first place. Thus, it is 'easier' to obey the exhortations of advertisers to eat certain types of unhealthy food or to drink a great deal of alcohol, when we know that medical means exist to relieve the symptoms these may cause. However, this can then lead to greater damage (e.g. ulcers or liver disease) which will have to be treated in more radical and possibly dangerous ways. In other words, the existence of effective symptom relief provides a rationalisation for not correcting the problem at source – a situation which is, of course, encouraged by all those companies who make a profit out of health-damaging commodities.

But perhaps the most serious form of iatrogenesis can be seen in the huge volume of psychotropic or mood-changing drugs prescribed each year (see chapter 4). Not only can these be health-damaging in themselves but they can numb the pain to the point where the sufferer will not act to change an unhealthy living or working situation but will continue to take the pills.[27]

Medical priorities

All these problems with medicine have been highlighted in recent years by its staggering cost. In 1982 over 10 per cent of the gross national product of the USA was spent on health care. Even in the UK, where socialised medicine provides much better value for money, the cost in 1982 was still about 6 per cent of GNP.[28] Expenditure on such a scale makes it essential that medical priorities should be rigorously assessed to ensure their cost-effectiveness, and it is open to considerable doubt whether current patterns of resource allocation encourage the provision of a humane and effective health care system. Considered from this standpoint, it is clear that much high-technology medicine does not justify the enormous public investment demanded by its powerful political lobby.

As an illustration, let us take two of the diseases to which we have already referred – cancer and heart disease. One of the most expensive pieces of medical technology developed in recent years is the 'whole body scanner' – a machine enabling a three-dimensional non-invasive inspection of the entire interior of the body. Each machine now costs close on a million pounds, and they are advertised as essential for effective cancer treatment. They are so expensive that most health authorities cannot afford them and they are frequently the subject of voluntary appeals for funds from the general public. However, there is little evidence to suggest that they aid the survival chances of those on whom they are employed.[29] Similarly, large sums have been spent equipping hospitals with intensive care units specialising in the treatment of coronary heart disease. Again, it is not clear that such treatment improves recovery prospects. Indeed, there is evidence to suggest that sudden entry into a highly technological and alien environment at a moment of biological crisis, may so disturb the patient as to sometimes lead to an actual deterioration.[30]

This question of the impact of the clinical situation on the emotions and feelings of the patient brings us to our final criticism of Western scientific medicine. Knowledge is now fragmented into specialisations which are practised by experts whom we have no choice but to trust for whatever services we may require. Among other things, this leads to compliance in the face of medical authority which, as we have seen, was an important feature of the development and application of the biomedical model. The mechanistic approach of medicine, its tendency to objectify human beings and its narrow focus on bodily symptoms, are often combined in the practice of an individual doctor, making the patient feel powerless, dehumanised and without resources. However, many more people are now complaining about being treated as objects, and feminists in particular have formulated fundamental criticisms of both the knowledge base and the social relations involved in the production of medical care.[31]

We have seen then, that modern medicine has been extensively criticised in recent years. For some people, this has led to political campaigns – thus the women's health movement has fought hard to change many of the most inadequate and demeaning aspects of current medical practice. Others, however, have turned to what is often called 'alternative medicine' in an attempt to find a replacement for the biomedical model which offers a better way of curing illness and/or promoting health.

Holistic alternatives to the biomedical model

The approaches to health and healing currently being adopted are often referred to collectively as 'holistic' or 'alternative' medicine. However such terms cover a wide range of differing practices; many have their roots in non-Western traditions (Chinese and Indian in particular), while others such as biofeedback and some types of meditation are compatible with, if not dependent upon, biomedical theories. Some define themselves as clearly antagonistic to Western scientific medicine, while others exist on its fringes, wishing not to challenge orthodox therapies but to extend them. But whatever their differences, all have in common the belief that there is more to maintaining health than the chemical or surgical control of physiological malfunction. They therefore seek to expand existing definitions of both health and healing, their rationale being aptly summarised by Abraham Maslow's comment; 'If the only tool you have is a hammer, you tend to treat everything as if it were a nail.' Holistic practitioners may practise techniques ranging from massage to homoeopathy, acupuncture, spiritual healing, herbalism and radionics – to name but a few. However, most hold, more or less, to a common set of basic assumptions which can be compared to the summary of the biomedical model given earlier.[32]

(a) Health is a positive state of being and not merely the absence of symptoms or disease.

(b) The precondition for health is the integration of the mind and body of the individual.

(c) When disease occurs, this is because of a disharmony between the individual and her/his environment or a lack of balance between the different aspects making up the individual human being.

(d) The precise nature of the disharmony will vary from person to person and will require diagnosis based on detailed knowledge of the individual concerned.

(e) Therapeutic emphasis should be placed on the capacity of persons to heal themselves. For this reason, treatment is addressed more toward creating a physical/psychological environment in which this capacity can express itself rather than attempts to *intervene* in the bodily process (e.g. drugs or surgery).

While specific alternatives may diverge from our typology, it is accurate enough to enable a preliminary assessment of most holistic practices. Their most obvious value is that – in theory at least – they extend the biomedical model to include individual consciousness as a key determinant of health. This is clearly important at both a practical and also a more philosophical level. We have already seen that psychological factors can be influential – both in the initial contraction of a disease and in the potential that exists for a cure. They should therefore be reflected in healing practices if such practices are to be successful. In more philosophical terms, the 'holistic' emphasis on the close relationship between mind and body can play an important part in broadening definitions of health itself.

In recent years, there has been a growing recognition that our current patterns of social organisation have led to a one-dimensionality and shallowness in people's perceptions of themselves, their work and each other, and to a weak sense of identity and an incapacity for self-fulfilment. So far these insights have not been incorporated into institutional conceptions of health and to the extent that holistic approaches link health with the quality of personal life ('happiness') then they are of potential value. They can provide a counterweight to the crude mechanism of Western scientific medicine while helping to democratise medical practice through encouraging people to understand how to look after themselves without becoming too dependent on doctors.

Yet on the other hand, holistic alternatives can lead to a process of 'victim-blaming' where people are held responsible for their own illness. This is because they emphasise the integration of mind and body but often pay too little attention to the social and economic context within which these 'integrated beings' live (or die). When wider environmental questions *are* taken into account, holistic approaches tend to focus on people taking control of their own *particular* environments, often ignoring the social and economic constraints on individual choice. In the first place, people are constrained by a whole set of social and economic expectations – those associated with sex roles for instance. Thus a woman who feels her mental health is being damaged by staying at home with young children might have little alternative but to continue. Furthermore, people with limited material resources will be particularly susceptible to advertising. Since they have few alternative sources of satisfaction many will continue to behave in ways that threaten their health – smoking cigarettes, for example. Finally, most people are faced daily with hazards about which they as individuals can do very little – dangers at work or unhealthy housing, for example. Thus to hold individuals entirely responsible fo their own health can detract attention away from the many aspects of society itself that are damaging to health.[33]

The relatively affluent are, of course, in the best position to control their personal environment and not surprisingly it is they who have benefited most from the holistic health movement, especially in the USA where alternative practitioners are at their most numerous and powerful. Yet for most people lack of education, money and confidence constitute a formidable barrier to exploring such alternatives.

This does not of course apply in most underdeveloped countries where so-called 'alternatives' will often be the cheapest care available.

It is important then, to formulate our criticisms of the biomedical model in ways that are constructive and take the needs of the entire population into account. Whatever their potential merits, holistic approaches often fail to do this. At a time when politicians are only too happy to find rationalisations for reducing the scope of the NHS, holistic arguments can sometimes provide hostages to fortune (as of course can other 'qualitative' critiques such as that coming from feminists). A growing number of people continue to require the curing and caring services that health services are able to provide. However, if they are not presented properly, holistic notions of individual responsibility and generalised critiques of the efficiency of orthodox medicine can only too easily be transmuted into justifications for cutting services.[34]

Any discussion of the potential of holistic medicine must at some point consider the important but difficult question of the assessment of its therapeutic efficacy. On the basis of existing knowledge, it would seem that the different types of alternative medicine vary enormously in terms of the predictability of their curative or preventive powers. Some have been assessed under controlled conditions while others have not, and those that have been tested have shown varying degrees of success. Acupuncture, for example, stands up relatively well to empirical assessment and appears to be particularly successful in a range of (often painful) conditions where conventional medicine has little to offer.[35] On the other hand, there is not much evidence that, say, homoeopathy works reliably – at least not in the way its proponents claim.[36] There does not seem to be a set of observable consequences which can be shown regularly to follow application of this method to particular cases. It is crucial that testing of this kind should be done and that it should occur under conditions designed to eliminate the so-called 'placebo effect' – the impact of either the healer's or the patient's belief in a therapy and its effectiveness.[37] Of course, this is not to say that such beliefs are not an extremely important part of the therapeutic process.

Alternative practitioners sometimes respond to these criticisms by arguing that their theories cannot be judged by a system of 'rationality' which is geared to the very 'science' which they wish to reject, and as a result they tend to rely on anecdotal evidence of success.[38] The issues behind this debate are complicated but two important points need to be made.

First, it is clear that the method of controlled trials is the same as would be applied in determining the technical success or failure of any form of human intervention in nature, from the building of bridges to the cooking of the perfect soufflé. Why then should they not be applied to any technique designed to improve human health?[39]

Secondly, we need to distinguish in this case between scientific method and scientific explanation. While it is certainly true to claim that Western scientific medicine may not have the conceptual apparatus to *explain* any successes holistic practices may achieve, this does not mean that the methods of empirical assessment associated with science should not be used to evaluate the efficacy of such practices – provided of course that criteria of success can be agreed upon. Indeed, it is important to emphasise that the scientific *assessment* of a treatment is logically separate, whether or not the method happens to have been developed from 'science'. Some drugs are arrived at 'scientifically', others empirically or by chance. Thus 'scientific trials' may also be used to validate *any* treatment that science cannot (yet) explain.

There will of course continue to be debates about these issues, but it would be strange if the alternative practitioner did not want to know whether or not her/his prescribed treatment did consistently prevent or cure an illness and whether or not its success or failure could be accounted for with the help of biomedical research. The connection between the success of acupuncture therapy, for example, and the stimulation of anaesthetising 'endorphins' in the brain is an interesting case of how the two can work hand in hand. Indeed, much of Chinese medical practice entails a pragmatic approach to healing, drawing on both traditional and Western techniques to great advantage.[40] Similarly, this compatibility is illustrated by the promise of 'relaxation therapy' in the treatment of stress through the use of biofeedback machines. These electronically represent internal bodily functions, so that patients can learn to exert conscious control over them, and their design is predicted on physiological measures of 'normality' employed in the biomedical approach (e.g. blood pressure, respiration rate, and so on).[41]

But whatever one's view of the ultimate compatibility of 'orthodox' and 'unorthodox' medicine, it is clear that the issue of predictable efficacy – however theoretically construed – is a vital one for potential patients. Many people turn to alternative medicine through desperation at the failure of orthodox approaches and often invest much of their remaining psychological and financial resources in the hope of a cure. Some form of empirical testing and the publicising of results is therefore essential if patients are not to be mystified and are to make rational and informed choices between alternative therapies. Naturally, this should also be applied to 'orthodox' medical techniques, since, as we have seen, they are by no means always as adequately tested as they might be, and few attempts are made to spread information about their efficacy to the general public.

The political ecology of health

There is clearly a need for a re-evaluation of 'health' and 'health care' in ways which can be empirically assessed and which do not ignore the value of certain aspects of the biomedical model. Let us call this re-evaluation 'the political ecology of health' and divide it into three components.

First, it will entail a shift in emphasis away from biological concepts of normality and physical fitness as the main indicators of health, towards an approach that maximises the individual's freedom to 'flourish' and to explore her/his individual potential. This should apply whatever the physiological state of the person concerned.

Thus questions of fitness are not irrelevant to the assessment of health but they must be directed towards a much wider selection of goals and expanded to include both psychological and social considerations. The question of the 'health' of the elderly or the physically disabled, for instance, is distinctly problematic when looked at in terms of conventional medical definitions. However, thinking in these new terms underlines the fact that effective treatment would have to go beyond conventional care to encourage them to act as creatively as possible – given their physical and mental potential.

Similarly, it is possible within the biomedical model to regard someone as 'healthy' who, while fit enough to maintain a job, may be wasting away in both emotional and intellectual ways. With a new and expanded definition of health, such a person would be seen as being in need of help, support and possibly a change in living and working conditions. In this way, the struggle for health can become a part of the general struggle for human freedom – for the abolition of those arbitrary constraints which keep us from discovering who we are through what we can accomplish.[42]

Second, the ecological dimension of this redefinition will involve a shift in emphasis away from individual pathology towards a concern with those aspects of the general environment which are disease-inducing or, conversely, health-promoting. As we have seen, one problem with the biomedical model is that it restricts the concept of treatment to the diseased organism. An ecologically-based theory of health would do the opposite through emphasising the extent to which environmental factors are responsible for disease. Furthermore, it would suggest ways in which the environment could be 'treated', rather than the patient.

There is of course nothing new about health and medicine considered in this way – at least in principle. Public health programmes in both the developed and underdeveloped parts of the world already work on roughly this basis. What would be new would be the emphasis placed on a critical reevaluation of whether or not we really *need* those products which are dangerous to produce or consume. Moreover, such an assessment would need to take place within a broader ecological context since many production processes, while not immediately damaging to individuals in the short term, may damage the environment in ways which will return to haunt us.[43] As an example of this circularity in practice, we could take the case of certain pesticides where the producer and farmer can both be damaged in the long term – possibly through contracting cancer – while the product itself then gets into the food chain and returns to destroy the health of consumers.[44]

There is no inherent conflict between our suggested re-evaluation and the understanding and techniques that are already a part of Western medical research at its most successful. There is however, a problem about the way such research is currently practised and the narrowness of its overall conception of environmental change – the way 'the environment' is defined in traditional epidemiology, for example. The first step towards a more radical epidemiology would be to focus on a much broader range of explanatory variables. This would include a closer examination of those conventional dangers which have already been given some consideration but it would also involve an investigation of the more qualitative hazards related to people's subjective experience of their own working and living environments.

This need for more subjective or experiential information brings us to the issue of how such research should be practised and by whom. It will inevitably involve the use of conventional scientific techniques associated with more traditional epidemiology and the participation of relevant 'experts'. However, at all stages, research must also include the active participation of the populations under investigation. Ultimately, the people involved are also *real* experts about the conditions under which they work, live, become ill and die, and this must be reflected in the research process itself.[45]

Third, and finally, the ecology of health cannot be divorced from wider political issues. Traditional epidemiological methods can provide only a very partial explanation of the social causes of ill-health and certainly cannot account for *why* particular health hazards exist in the form that they do. This is because they cannot explain the links between these different hazards and the wider social and economic system in which they are found. Recognition of the importance of these issues lies at the heart of a radical epidemiology. We have already examined the historical relationship between orthodox medicine and the development of industrial capitalism, and there is abundant evidence that much contemporary illness is inextricably linked to capitalist relations of production.

It is therefore misleading to refer to such dangers in ways which underplay their social and economic origins. Such an approach tends to 'medicalise' epidemiological research – orienting it only towards the immediate environmental causes of illness and the introduction of protective measures that do not deal with the problem at source (e.g. better protective clothing for individual workers or lower emission rates for toxic waste from individual factories). While it is clearly important to criticise the biomedical model for its neglect of environmental disease causation, we need to go much further in our critique. We need to show how traditional epidemiology tends to focus either on the individual or on immediate isolated hazards, thus drawing attention away from more underlying problems within the social system itself which depends for its survival upon the violence it does to the environment and to those who produce wealth within it.[46]

To conclude then, a political ecology of health would emphasise the need for changes in the way individuals are treated within the medical care system. It would place a much greater stress on the subjective feelings and the essential 'humanity' of those in need, whilst not denying them the real advantages of technological medicine. The treatment given to a sick person might well be of the kind now characterised as 'scientific' but it could also involve whichever of the 'alternative' therapies had been shown by rigorous testing to be the most appropriate for their particular needs.

A political/ecological approach would also emphasise the environmental causes of ill-health. It would make a scientific analysis of the ways in which our present mode of social and economic organisation damages both our physical and mental health. From this analysis it should then be possible to begin the formulation of what a healthier society might look like – and how we might go about achieving it.

6. Human sociobiology

BSSRS Sociobiology Group

Sociobiology has been defined as 'the systematic study of the biological basis of all forms of social behaviour, in both animals and human beings'. Widespread debate about human sociobiology dates from 1975 and the publication of E. O. Wilson's book, *Sociobiology: The New Synthesis*.[1] In this book Wilson spent twenty-six chapters considering the application of sociobiological ideas to the understanding of animal behaviour. Then, in a final chapter, entitled 'Man: from sociobiology to sociology', he attempted to demonstrate the utility of evolutionary theory to understanding human social behaviour and social organisation. Enormous controversy, especially in the USA, has been generated by this chapter and by similar work which has followed it. Wilson himself added to this in 1978 when he published *On Human Nature*,[2] a full length text entirely given over to human sociobiology.

It is not, of course, novel to try to explain human behaviour within a biological framework. The Victorian upper classes, for example, responded favourably to social Darwinism, and its suggestion that their social superiority reflected their biological superiority. The poor were poor because they were unfit and no amount of social reform could change this underlying, biologically-based, inferiority. The eugenics movement had its origins in the same beliefs. Eugenicists advocated the sterilisation of 'undesirable' social elements because they believed this undesirability to be biologically-based and they did not want it further reproduced. More recently, debates about IQ heritability and race have drawn on the same type of idea.

These ideas have always found a popular audience too. In the UK, Desmond Morris and Robert Ardrey are amongst the best known popularisers, and Morris' first book, *The Naked Ape*, has sold over eight million copies with its 'sports cars are really penises' type of analysis.[3]

The new human sociobiologists try to set themselves apart from these predecessors however. They argue that their approach is more rigorous than previous work, and that it has a more substantial intellectual foundation because it is based on contemporary developments in evolutionary theory. Academic human sociobiology suggests that there is, or soon will be, a 'hard' theory which will support both common prejudice and the popular science of writers like Morris and Ardrey. It promises to demonstrate that we are all *biologically* designed to be individualistic, competitive, strongly sex differentiated, and all other characteristics that Western society assumes to be universal.

This chapter begins by summarising the main developments in modern evolutionary theory upon which sociobiology draws, then goes on to identify and critically evaluate a number of its key assumptions. After arguing that such a theory has little to offer in explanation of human social behaviour, consideration is given to its contemporary political and ideological significance. In conclusion, there is an outline of an alternative theory of human evolution: a theory which gives prominence to the unique characteristics of human beings.

Contemporary developments in evolutionary theory

Biologists seek to explain animal behaviour by showing how it is adaptive; that is, how it contributes to the individual animal's fitness. Applying this approach to social behaviour however, raises an immediate difficulty. This difficulty is how to explain altruistic behaviour. When an individual animal gives an alarm call it attracts attention to itself and increases its risk of attack by the predator. Giving an alarm call thus reduces individual fitness.

The difficulty of explaining this behaviour in terms of the maximisation of individual fitness led some biologists to suggest that natural selection operates at the level of the group rather than the individual. Alarm calls protect the group and thus have adaptive significance in terms of the group, if not of the individual callers.

While initially attractive, group selection arguments cannot account for the continued existence of altruistic behaviour. The argument works only as long as it is assumed that every member of the group has evolved with the same propensity to give alarm calls. If, at any time, individuals appear with a reduced propensity then altruistic behaviour will eventually disappear because the individuals who do not give calls, and who do not risk themselves, thrive at the expense of those who do. Since it is unlikely that such differing propensities would not have evolved during evolution and since alarm calling is a well established behaviour, then group selection arguments fall.

Contemporary biology has developed a number of approaches to explain how altruistic behaviour could have evolved. One of the most powerful is kin selection.[4] Kin selection proposes that genes rather than individuals are the unit of natural selection. Individual animals are concerned to maximise the survival of their genes. Importantly, these genes are carried not just by the animal itself; some are also shared with relatives, and the closer the relative the more genes which are shared. Whilst altruistic behaviour may lower individual fitness, at the same time it may increase *inclusive fitness*, that is the combination of an individual's reproductive success *and* that of its kin. 'Altruistic' behaviour between related individuals is thus consistent with a theory of kin selection.

Human sociobiology

> The question of interest is no longer whether human behaviour is genetically determined; it is to what extent. The accumulated evidence for a large hereditary component is more detailed and more compelling than most persons, including even geneticists, realise. I will go further: it is already decisive.[5]

So argues E. O. Wilson in *On Human Nature*. For him and

for other human sociobiologists our social behaviour is as much a product of natural selection as our opposable thumb. To such commentators human nature is natural.

They draw on a range of information to support their argument. David Barash, a well known sociobiologist, has pointed to the enormous importance of family and kinship networks in human societies, for example. He notes that anthropologists have experienced difficulty in agreeing why kinship is such a universally important organising principle and suggests that an evolutionary perspective can provide an answer. Kin selection, he believes, suggests that kinship would be of paramount importance in human societies; he thinks it also predicts nepotism and hostility to strangers.

> We help those in whom we recognise ourselves and we do so in proportion as we (or more precisely our genes) are represented in others. Such is the fundamental principle of nepotism, one of the most pervasive forms of behaviour in higher vertebrates, and most notably in our own species.[6]

We help our relatives because they are our relatives. In helping them we are helping ourselves (our genes). As genetic relatedness becomes more distant so hostility increases; hence the fear of strangers.

Explaining sex differences in the distribution of power and of social roles has provided another major focus for human sociobiology. The argument begins from what sociobiologists see as a fundamental asymmetry between the sexes. Richard Dawkins describes it thus:

> There is one fundamental feature of the sexes which can be used to label males as males and females as females, throughout animals and plants. This is that the sex cells or 'gametes' of males are much smaller and more numerous than the gametes of females . . . *it is possible to interpret all the other differences between the sexes* as stemming from this one basic difference [emphasis added].[7]

The argument is that although females and males contribute equal numbers of their genes to their offspring, the female commits more resources with her food-rich ova. Producing them involves her in the expenditure of more resources than the male who can produce large numbers of low-resource sperm. The consequence is that:

> he is able to beget a very large number of children in a very short period of time using different females. This is only possible because each embryo is endowed with adequate food by the mother in each case. This therefore places a limit on the number of children a female can have but the number of children a male can have is virtually unlimited. *Female exploitation begins here* [emphasis added].[8]

Internal fertilisation, a fairly long gestation, and lactation make the difference between female and male reproductive potential especially acute in mammals. Daly and Wilson (a different Wilson), in a student text on sociobiology, contrast the woman credited by *The Guiness Book of Records* with a record of 69 live births with the male who fathered a record of 888 children.

From this difference in reproductive potential, and drawing on the theory of kin selection, human sociobiologists have developed a number of arguments about human social behaviour and social organisation. They believe that using these ideas they can explain human sex differences in parental investment, male selection, courtship strategies, sexual codes and sexual conduct.

Trivers has defined *parental investment* as 'any investment by the parent in an individual offspring that increases the offspring's chances of surviving . . . at the cost of the parent's ability to invest in other offspring'.[9] In one sense, it is in neither sex's interest to parent; it is much better for each of them to persuade the other to carry the major, if not the whole, responsibility. In mammals, however, because of internal fertilisation, long gestation and lactation, the female ends up with prime responsibility for the care of offspring. This adds up to a much greater commitment and allows the male to 'invest' in offspring with other females.

That is why in human societies, the sociobiologists argue, child rearing is a female responsibility.[10]

They use a similar argument to explain why *polygamy* is a more common social arrangement than either monogamy or polyandry. Male mammals can mate with several females in quick succession, without detriment to the survival of the offspring, so long as females bear the main burden of parenting. Greater *male promiscuity*, or willingness both to initiate and have sex (at least in the cultural assumptions of many societies) follow from the same logic. Even where the female and male form a pair-bond (monogamous couple) the male may nevertheless seek chances to impregnate other females.

The different parental investments of females and males mean that they require different qualities in a potential partner. In selecting a mate, females want good providers whilst males want good breeders (to put it crudely). Human sociobiologists believe that they see evidence of this in human society in the practice of women marrying men older than themselves. An older man will have shown whether he will be a good provider, a younger women has all her reproductive years ahead of her.[11] And the data on sexual attraction which suggest that 'a woman's desirability . . . depends upon her physical attractiveness to a greater extent than does a man's . . . his worth as a potential mate depends somewhat more upon his status and prospects for success',[12] reinforce their point.

Human sociobiologists make much out of the so-called *sexual double standard* whereby adultery/promiscuity are tolerated in men but not in women. Barash, for example, notes that:

> The philandering husband is no great disgrace, as long as he provides adequately for his domestic obligations. The cuckolded husband, on the other hand, is an object of ridicule . . . Sociobiology predicts that human males are . . . significantly more intolerant of infidelity by their wives than wives are of their husbands . . . (female) adultery is punished with particular severity.[13]

Sociobiology makes this prediction because, while everyone knows who the mother of a child is, only the mother can be sure who the father is. Males must therefore police female sexual activity in order to maximise the likelihood that *they* will sire the children of their partner and not waste parental effort on another man's child. Hence the emphasis on female virginity and fidelity, and hence too, the practice of patrilocality. This refers to women moving to their husband's homes/villages on marriage, the sociobiological reasoning being that this allows the in-laws to protect their interests more carefully.

Finally, the supposedly greater *sexual assertiveness* of the human male is suggested by sociobiologists to be a consequence of inter-male competition for females. For males in both human and animal species, 'the male who wins the right to inseminate a female also wins for his progeny a share of the female's potential parental investment'.[14] Human sociobiologists recognise that there are cultural variations in all of these behaviours and practices but they maintain that there is a sufficiently clear pattern of similarities across cultures for this to be evidence of a common basis.

A critique of human sociobiology

In the years since E. O. Wilson's famous last chapter was published, a number of important critiques of human sociobiology have been published.[15] Rather than reviewing all of these we offer here a critical analysis of its underlying assumptions and assertions. The following notions are challenged:

(a) that if a behaviour is universal then it is likely to be under genetic control;

(b) that cultural variation and diversity is, in the end, superficial and not a significant factor in the explanation of human social behaviour;

(c) that if a behaviour can be said to have biological advantage then it is likely to be genetic in origin;

(d) that if a behaviour has social advantage then it has biological advantage too;

(e) that human society can be explained in terms of the aggregation of individual characteristics;

(f) (perhaps most important of all) that studies of animal behaviour provide a legitimate source of information about human behaviour.

If a behaviour is universal then it is likely to be under genetic control

The initial issue which arises here is whether there are universal social behaviours. Anthropologists and sociologists have traditionally been reluctant to talk about universals of human behaviour irrespective of meaning, context and intention. Sociobiologists do not share their caution. E. O. Wilson, Barash, van den Berghe, and Daly and Wilson, all make grand and confident statements about universals in human societies without any reference, or with only the most cursory reference, to meaning, context and intention.

They present universal behaviours as uncontroversial, but all too often what they have done is to rummage selectively through the anthropological and sociological literature picking out those items that suit their arguments. They ignore the often complex and controversial arguments over interpretation. For example, E. O. Wilson in his recent book with C. Lumsden, *Genes, Mind and Culture*, confidently assures the reader that in mother–child interaction, 'close contact between the mother and her infant during the first hours following birth appears to be crucial for the formation of subsequent strong bonding'.[16] However, a number of recent studies[17] do not support this assertion, and some believe the original results on which they were based are an artefact of the experimental procedures used.[18] Lumsden and Wilson do not mention this, preferring instead to cite mother–infant bonding as a prime example of epigenetic rules functioning 'independently of the cultural surround or any learning experience'. Yet there is plenty of evidence suggesting that the quality of early mother–infant interaction is intimately bound up with the cultural 'surround'.[19]

Faced with challenging counter-evidence sociobiologists

resort to *ad hoc* arguments. As we have seen, sociobiological theory predicts that males will woo females and that females will select among males. In modern Western societies, however, males select females; and Levi-Strauss, the French anthropologist, has even suggested that the exchange or barter of women is a key, if not the key, facet of human social organisation. Undeterred, van den Berghe and Barash tell us that while it may *look* as if men select/exchange/barter women,

> this public official, legal display is not necessarily an accurate reflection of how the marriage came about. Women *are* in fact often consulted, and may even play an active behind-the-scene role . . . Even in our own society, for instance, there is a fiction that the father of the bride 'gives her away', but *we know better*. The point is that, in all societies, rape is the exception rather than the rule. Notwithstanding male dominance, female assent is a pre-condition to nearly all mating [emphasis added].[20]

Here marriage and mating become equivalent. Because, in van den Berghe and Barash's view, most women submit to intercourse without too great a struggle, this constitutes the equivalent of females choosing mates, and their theory survives. The vast anthropological and sociological literature on the subordination of women counts for nothing: behind the scenes, detected by van den Berghe and Barash's intuition, women – universally – are doing the selection. So much for scientific rigour.

Their arguments about universality are also, on occasion, inconsistent. Van den Berghe and Barash, for example, offer the following strong statement about human polygamy:

> Our species tends to show a preference for limited polygamy. Some three-fourths of all human societies permit polygamy and most of them prefer it. Monogamous societies often have been polygamous in a more or less recent past, and typically their monogamy is a legal fiction. That is, legal monogamy is combined with various forms of

> institutionalised concubinage. Polyandry (where a woman has multiple male partners) on the other hand, is *extremely rare* and linked to very special conditions [emphasis added].[21]

Three pages later though, we are told,

> It is true, of course, that social advantages of wealth, power or rank need not, indeed often do not, coincide with physical superiority. Women, *in all societies*, have found a way of resolving this dilemma by marrying wealthy and powerful men while taking young and attractive ones as lovers [emphasis added].[22]

Faced with such data it is difficult even to accept that there are universal behaviours, never mind to discuss whether or not such behaviours are under genetic control.

There is however, one human behaviour which almost everyone agrees comes close to being universal and this is incest avoidance. Few anthropologists and sociologists would disagree that some form of incest taboo is found in almost every society. Sociobiologists suggest that there is a simple reason for incest avoidance – biological advantage.

Prima facie, incest avoidance is an excellent candidate for a human social behaviour that is genetically based. There are very strong biological advantages in not mating with close relatives, and so it is a behaviour that could have been the subject of natural selection. Incest avoidance probably provides sociobiologists with their strongest apparent instance of a genetically-based human social behaviour. It is therefore all the more interesting to examine it carefully.

The detailed prescripts of people with whom sexual intercourse is allowed or forbidden vary between societies, and in some non-kin are included as well as kin. Mostly, however, intercourse is forbidden between parents and their offspring and between siblings. Data on the frequency of incest are very difficult to generate; it is a taboo subject in more than one way. In both the UK and the USA however, recent data suggests that its incidence is far greater than previously recorded.[23]

This raises difficulties for human sociobiology. Let us assume, for a moment, that incest avoidance has a genetic base, that there is, in Barash's words, a 'whispering within' which tells us to avoid it. The likely incidence of incest in the UK and the USA suggests that the whisper cannot be very loud. If incest occurs on any scale *despite* strong biological disadvantages *and* cultural taboos, then what is the likelihood of the genetic control of other behaviours for which the sociobiologist's case is weaker?

It is also unnecessarily generous to grant that incest has a genetic base. Merely because a human behaviour is common is not sufficient reason to suggest that it is genetic. Religious observances are widespread in human cultures: are they genetic?

Cultural variation and diversity is, in the end, superficial and not a significant factor in the explanation of human social behaviour

Cultural diversity is dismissed by the sociobiologists. Barash argues, for example:

> We human beings like to think that we are different. We introspect, we are confident that we know what we are doing and why. But we may have to open our minds and admit the possibility that our need to maximise our fitness may be whispering somewhere deep within us and that, know it or not, most of the time we are heeding these whisperings.[24]

E. O. Wilson deals with cultural diversity by positing a multiplier effect: small genetic differences can be amplified by culture to large cultural difference. In their turn, Daly and Wilson argue that 'the denial of biological constraints upon our behaviour amounts to an absurd assertion of chaos.[25] For them cultural diversity is a product of ecological differences and of differences in gene pools, and they refer to 'arbitrary cultures'.

It is however crude and naive both to dismiss cultural variation as behaviourally insignificant and to regard it as arbitrary. Take variations in monogamy and polygamy as an example. Mervin Harris, an anthropologist, has made the following observations:

> Human sexual behaviour is . . . so diverse as to deny any species-specific characterisation. The heterosexual range runs from promiscuity through monogamy, with each type practised by tens of millions of peoples. Human females in general may not have plural mates as often as males, but there are millions of women who obtain a plurality of sexual partners as often as, if not more often than, the most active men in other societies . . . The idea that males naturally desire a plurality of sexual experiences while women are satisfied by one mate at a time is entirely a product of the political-economic domination males have exerted over women as part of the culturally created, warfare-related male supremacy complex. Sexually adventurous women are severely punished in male-dominated cultures. Wherever women have enjoyed independent wealth and power, however, they have sought to fulfil themselves sexually with multiple males with no less vigour than males in comparable situations.[26]

Harris is suggesting that women's pattern of sexual behaviour is related to their economic and political position; in his account economic and political factors *structure* cultural variation. Variation, consequently, is not arbitrary. An analysis by Karen Sachs of the incidence of extramarital sex amongst African tribes further illustrates this point:

> Mbuti and Lovedu have a single standard regarding extramarital sexual affairs. Pondo women view their extramarital affairs as right and proper, but the men see women's affairs as immoral. A Ganda husband may kill his wife for real or suspected adultery, but a wife has little recourse against her husband.

This reflects the economic differences between the tribes:

> In Mbuti and Lovedu both sexes perform social labour in a use economy; among the Pondo, this remains the organisation of women's labour, but the men perform social labour at least in part in an exchange economy; and in Ganda women's work is individual domestic production for household use, while men work in groups, almost totally in production for exchange.[27]

Patterns of sexual behaviour then, are neither universal nor subject to arbitrary cultural variation. Rather they follow from patterns of social, economic, and political relationships.

If a behaviour can be said to have biological advantage then it is likely to be genetic in origin

We have already seen with the example of the incest taboo that because a behaviour is biologically advantageous is not a sufficient reason to suggest that it is genetic. It was biologically advantageous to attend Mass when the Inquisition was in town, yet no one would suggest that attending Mass has a genetic basis! In the anthropological literature there are many discussions of people behaving in ways that are biologically advantageous to them, such as observing certain food taboos; but anthropologists do not assume that such taboos are adopted because of genetic control. Indeed, there is a school of anthropology that explains many cultural practices that are biologically adaptive, from kinship patterns to prohibitions on pork eating, as learned behaviours.[28] Given that such theories have been elaborated precisely to explain why the behaviour that people adopt coincides with behaviour that is advantageous biologically, it is naive of sociobiologists simply to take for granted that any piece of biologically advantageous behaviour must be under genetic control.

Even if we find a behaviour that is biologically advantageous and also appears to involve a genetic component, the connection between the two may be far more complex than sociobiologists allow. Consider, for example, Mead's discussion of the taboo in hunting societies on men being present during a birth. She believes that everyone has an innate propensity to wish to stay with and care for babies they have seen born. This would clearly cause problems if a society relied on men going on long hunting trips and they refused to go. Therefore, she argues, a taboo has grown up against men watching the birth. In this type of model, genetic factors are not denied but they are interrelated in a very complex way with cultural factors and are not themselves the determinants of biological advantage.[29]

If a behaviour has social advantage then it has biological advantage too:
Sociobiologists sometimes rest their claims about biological advantage on nothing more than evidence of social advantage. They assume that a characteristic is biologically advantageous and thereby increases fitness, when all they have done is to point vaguely at some social advantage that it brings. It may be that the two are connected in some cases. But this should be demonstrated, not merely assumed.

Take, for example, E. O. Wilson's discussion of Jamaican slave society. He cites it as an example of behaviour that was maladaptive and therefore on its way to extinction. Bear in mind that he is supposed to be providing evidence of *biological* maladaptation, so the examples should be of fitness-lowering behaviour. Here is the passage he quotes together with his comment on it:

> 'The slave society of Jamaica . . . was unquestionably pathological by the moral canons of civilised life. What marks it out is the astonishing neglect and distortion of almost every one of the basic prerequisites of normal human living. This was a society in which clergymen were the "most finished debauchees" in the land; in which the institution of marriage was officially condemned among both masters and slaves; in which the family was unthinkable to the vast majority of the population and promiscuity the norm; in which education was seen as an absolute waste of time and

> teachers shunned like the plague; in which the legal system was quite deliberately a travesty of anything that could be called justice; and in which all forms of refinements, of art, of folk ways, were either absent or in a state of total disintegration. Only a small proportion of whites, who monopolised almost all of the fertile land in the island, benefited from the system. And these, no sooner had they secured their fortunes, abandoned the land which the production of their own wealth had made unbearable to live in, for the comforts of the mother country.' Yet this Hobbesian world lasted for nearly two centuries.[30]

It may be that such features of a society as debauched clergymen and unjust laws lead to evolutionary disaster but it is by no means obvious. Wilson's list tells us a great deal about the social and moral characteristics of the society but he gives us no hint of what these might have to do with biological maladaptedness. History, unfortunately, is full of examples of societies which flourished, often for long periods, but which were cruel and unjust, and in E. O. Wilson's terms biologically maladaptive.

Human society can be explained in terms of the aggregation of individual characteristics

Put simply, sociobiologists assume that human society is nothing more than the sum of individual social behaviours. The assumption is that society can be explained largely in terms of the genetic characteristics of its individual members. The customs, institutions, events, and regularities that make up a society are merely a reflection of the dispositions, assumptions and beliefs of its members. The rationale for this view is that society is the product of interacting individuals, so social phenomena must ultimately be reducible to the composition and qualities of these individuals. If society is aggressive, selfish or competitive, then this must stem from our aggressive, selfish, competitive biological natures. Sociobiologists thus reduce very complex social phenomena and processes to a simple, individual level.

There is a great deal wrong with this position. Most

important, it ignores the fact that human beings are essentially social: our actions do not take place in a vacuum but within a social environment that influences them and the effects they have. We operate within a network of relations with others, and within institutional frameworks. The very complexity of our social environment, the ways in which our actions are mediated by social institutions and by the actions of others, all mean that individual behaviour can only be understood by reference to its social context.

Take queuing for example. British people form queues at shop counters because it is expected of them. Some cultures do not share this convention and do not queue. Social conventions also change with circumstances. Where mutual trust begins to weaken an orderly queue may suddenly break into a mad rush. Sociobiology is clearly of little use in helping us to understand even this mundane example of everyday social behaviour.

This does not stop sociobiologists confidently pronouncing the biological basis of far more complex social phenomena such as war. E. O. Wilson:

> Are human beings innately aggressive? This is a favourite question of college seminars and cocktail party conversation, and one that raises emotion in political ideologues of all stripes. The answer to it is yes. Throughout history, *warfare, representing only the most organised technique of aggression*, has been endemic to every form of society, from hunter-gatherer bands to industrial states. During the past three centuries a majority of the countries of Europe have been engaged in war during approximately half of all the years; few have ever seen a century of continuous peace [emphasis added].[31]

Wilson (and many 'proto-sociobiologists' such as Tinbergen and Lorenz) clearly see war as an outburst, en masse, of the 'aggressive instinct'. If this were so, why do soldiers often have to be conscripted, and made to fight? How often, when soldiers find themselves in the trenches, do they feel bored, uncomfortable, frightened and homesick rather than aggressive? Why do armies threaten certain death to the soldier who turns away from the front?

Even if people were innately aggressive this would still not explain, as sociobiologists seem to think, why we have wars. It does not account for the frequency of war changing dramatically between different countries and over time, while genes presumably remain much the same. For example there were only about thirty years between 1500 and 1700 in which there were no large-scale military operations between states. In the nineteenth century, on the other hand, conflicts were relatively infrequent. Sociobiologists have had little to say about such variability. But it is at the very centre of debate between historians, who find no need to invoke genetic imperatives. The historian Perry Anderson, for example, argues that in feudal times 'war was possibly the most rational and rapid single mode of expansion' available to the ruling nobility;[32] hence their obsession with war and valour. By contrast, in the nineteenth century capitalists replaced the lance with the punch-clock, and the charger with the steam engine. They made their wealth through the sweated labour of their factory employees and not through the capture of territory. So peace reigned (at least internationally) regardless of what genes had to say about the matter.

So, positing an innate aggressive instinct goes almost nowhere towards explaining the phenomenon of war. Nevertheless, the latest sociobiological treatise by Lumsden and Wilson extends the philosophy of innate predispositions to all aspects of society. They are adamant that society is nothing more than a conglomeration of individual instincts:

> Culture is in fact the product of vast numbers of choices by individual members of the society. Their decisions are constrained and biased [by epigenetic rules] in every principal category of cognition and behaviour thus far subjected to developmental analysis . . . laws governing culture qua culture must exist, but they can be synthesised from principles governing the mind.[33]

Their thesis, buried under a tangle of mathematical equations, is depressingly simple: if it feels right, people will do it. They see genes as setting in motion a compendium of epigenetic rules that structure our brain tissues so that some kind of ideas (which they call 'culturgens') are more thinkable and acceptable than others. When deciding how to behave individuals are supposed to 'consult . . . the oracle [sic] residing within the epigenetic rules'.[34] They allow that social conventions may 'amplify' epigenetic rules (as if culture were a hollow sounding box) but claim that in,

> several important cases, including brother-sister incest, patterns of colour naming and mother–infant bonding . . . epigenetic rules guide the assembly of behavioural programmes in a manner little influenced by the existence or nature of other culturgens.[35]

In making statements like this Lumsden and Wilson are assuming that culture can be broken into objective units or 'culturgens', but as we have already noted many social psychologists do not accept that there is an entity called 'mother–infant bonding'. Similarly our chapter on ethology (chapter 7) points out the reified nature of concepts like aggression, dominance and territoriality. That sociobiologists assume certain culturgens and not others testifies to the authors' own ethnocentrism – they live in the sort of society that values such concepts. The authors' political sympathies are surfacing.

They also imply that culture exists outside the individual and that we merely flip through its catalogue of possibilities picking and choosing. We have already shown that this is not the case. Choice is highly constrained – not by any genetic prescription, but by the structure of society.

Studies of animal behaviour provide a legitimate source of information about human behaviour

Sociobiologists make use of studies of animal behaviour to indicate parallels with, and to interpret, human behaviour.

Since animal behaviour is less controversially a product of natural selection, sociobiologists use animal–human parallels to support the 'human behaviour has genetic origins too' case. There are however a number of major problems associated with such an exercise.

An immediate problem is that the categories used to interpret animal social behaviour are derived from human behaviour. Donna Haraway has documented how this human–animal–human problem was initiated in the study of primates (a key comparative group for sociobiologists).[36] Given such a human–animal–human pattern it is hardly surprising that sociobiologists are able to discover so many animal–human behavioural parallels. Hierarchy, aggression, dominance, territoriality are all identified as organising principles of animal behaviour and extrapolated back to humans to become organising principles of our behaviour too. But, as chapter 7 shows, the concepts so derived are now under serious attack, even for understanding primate behaviour.

A second difficulty with such extrapolation is that many apparent similarities in behaviour may belie important differences in the function and meaning of these behaviours. Primatologists now recognise, for example, that the chimpanzee's broad 'smile' is better interpreted as a fear grin. It was not always recognised as such, and almost certainly, ethologists today are seriously misinterpreting other animal behaviours (see chapter 7).

Third, extrapolation relies on the observation and comparison of traits that are 'conservative' for primate species – traits that do not vary between species. Even sociobiologists recognise that there are problems in this approach. Language, for example, would not be considered genetically-based on this approach. Yet from neurological evidence it seems probable that the capacity for language *has* a genetic component. It is absence of language, however, that is a conservative trait amongst primates. In which cases is it reasonable to assume a sharp break, or order of magnitude difference, and in which cases not? Sociobiologists do not address this question or offer any criteria to assist the distinction.

The human animal is unique

We believe that there are a number of significant differences between humans and other animals. These differences are important for understanding and interpreting human social behaviour, and are ignored by sociobiologists in their mono-causal and reductionist interpretations.

There are, first, a number of physiological/anatomical features – upright gait, opposable thumb, larynx – which are widely recognised as having made a major contribution to our social evolution. Our opposable thumb, for example, makes our hand a uniquely versatile, sensitive and flexible tool, allowing us a range of activities far beyond that of any other animal.

Second, there are a number of attributes that we share with other animals, but which are so much more developed in us that we are taken an order of magnitude further than other species. Our capacity for learning and for verbal communication are examples. While there is clear evidence of learning in animals, in humans both the proportion of learned behaviour and the capacity for learning are significantly greater. Even the most well-tutored ape or monkey is rapidly overtaken by the young human. The diversity of human behaviour across time and across cultures is testimony to our capacity for social learning.

We also have a very much more complex system of both verbal and non-verbal communication. Importantly, too, in many cultures we have amplified our system of verbal communication through the written and printed word. We have developed a system of cultural transmission based upon this which enables knowledge to be preserved and transmitted. The repository of knowledge consequently available to humans is far greater than the parent–child, peer–peer, superordinate–subordinate patterns of skill transmission found in other primates.

A third important difference is the human ability not only for separating the conception and execution of a task, but also for planning ahead. Some primates will co-operate in executing tasks – such as finding food – and they can solve problems. What they cannot do, which we can, is to plan this co-operation in advance, and to organise it in the sense that A formulates an activity for B and C to execute either now, or at some point in the future. These abilities for planning and executing tasks, for learning and organising them, for separating them over time and space, for dividing them between individuals, for communicating them via the spoken or written word (or more recently electronic signals) are fundamental characteristics of human beings which, as we shall see in the next section, underlie our status as a working, as *the* working animal. They make possible the labour process.

Fourth, unlike other animals we consciously create our history. Most of us spend most of our lives without recognising that the social order of which we are a part is a social construct which can be changed by our collective action, and changed in a way we choose. Very rapid social change – such as the events of the Chinese Revolution – can only be explained as the self-conscious acts of human beings. Such events have no parallel amongst animals.

Taken together we would argue that these abilities – learning, communication, abstract thought and reasoning, meaning and intent – make us human, and different, from other animals. They constitute a definitive challenge to the legitimacy of the animal–human extrapolations that sociobiologists so easily make.

The political and ideological significance of human sociobiology

We are not alone in our critique of human socio biology. Leach, reviewing Lumsden and Wilson's book remarked:

> This book comes so close to being a parody of the genre to which it belongs that I have difficulty in believing that it is not intended to be an academic hoax.[37]

And the reviewer in *Animal Behaviour* said:

> It does not provide new, clear primitive concepts; it reifies old ones. It does not provide new insight into dynamics; it assumes away, under the rug, some of the most important kinds of social dynamics.[38]

Why, then, has human sociobiology been so widely received in scientific circles and why has it found such a large popular audience?

The political impact of sociobiology is best understood if it is viewed as one of a line of explanations of the human condition that have grown out of science since the late nineteenth century. At that time science weakened the hold of the religious framework by which people understood and regulated their social interactions. Since then a number of doctrines explicitly claiming a scientific base have performed a related social function: social Darwinism in the late nineteenth century, eugenics at the beginning of the twentieth, behaviourism and psychoanalysis between the world wars, and, more recently, sociobiology.

Richard Dawkins, for example, insists that the welfare state is 'a very unnatural thing . . . inherently unstable because it is open to abuse by selfish individuals ready to exploit it'. There is nothing for instance, 'to stop a couple with no material resources at all having and rearing as many children as the woman can physically bear. Whereas in "nature" any gene for *overindulgence* is promptly punished: the children containing that gene starve [emphasis added].[39] It takes someone of a particular political persuasion to imply that the welfare state might encourage the spread of genes for "overindulgence". A class of people that is so exploited that it cannot afford to feed and clothe its own children is thus slighted as a parasite on the rest of us. It is perfectly right and natural, on the other hand, that the wealthy should go forth and multiply!

The power of these doctrines comes from three characteristics. First, as sets of ideas, they have a common two-part structure. A core of rather limited phenomena exist for which the doctrine provides quite a tight explanation – conditioning for behaviourism, dreams for psychoanalysis, the "altruistic" behaviour of the social insects – for sociobiology.

Then there is a vast range of everyday phenomena for which the doctrine provides a plausible, but extremely loose, explanation.

Second, because most of them provide a 'natural' explanation for the existence of social practices that appear unjust, they can be used to justify the practice. Thus social Darwinism was used to justify the fiercely competitive and exploitative practices of Victorian capitalism; eugenics was used in the early years of this century to argue against the usefulness of social reform – the condition of the poor, after all, came from their being 'bad stock'.

Third, their general concepts reflect a fundamental aspect of the social world in which people live and so are both easy to grasp and intuitively plausible. Thus social Darwinism, with its elevation of the survival of the fittest into a social doctrine, not only served as justification for the rapacious capitalism of the Victorian period but also was made plausible by it. Survival of the fittest was a principle that people could see in action all around them; it is easy to see why people believed that it was a fundamental principle of all social life. Behaviourism had as its central idea the notion that the springs of all action can be expressed as a single variable, namely the amount of reward and punishment such acts have previously produced. This was made plausible by the type of society in which it was created: the expanding capitalism of early twentieth-century USA where money was the measure of all things.

Behaviourism and sociobiology provide an interesting parallel. Behaviourism can now be viewed with the benefit of hindsight, and can thus provide a guide to how sociobiology might develop. Both began with influential papers – Hamilton's (1964) paper for sociobiology and Watson's (1913) paper for behaviourism. Both were novel in their approach and provided a technical method that could be carried out on simpler organisms – insects for sociobiology, rats for behaviourism. Moreover, both appeared *in principle* to be applicable to people. They had on their side fashion, apparent social relevance and, within circumscribed limits, an ability to generate new facts and mini-theories. Both too were reductionist; they provided the intellectual allure of taming the complexity of human interactions with a single tidy set of simple principles.

What happened to behaviourism typifies the historical trajectory of such doctrines. In its time it too dominated a whole scientific field, that of experimental psychology. But if one compares the ideas of Watson, the originator, with his most distinguished successor, Hull, in the 1940s, the original tidy set of principles had become a complicated set of *ad hoc* assumptions that attempted to account for an increasingly obscure set of phenomena. Eventually, by the 1950s, the theoretical edifice collapsed under the weight of the *ad hoc* assumptions that had to be made, and the triviality of the phenomena that were being studied by comparison with those that, in principle, the theory should have explained.

We can already see this process happening for sociobiology. For example, one reviewer of *Genes, Mind and Culture*, whilst welcoming the work, confesses to:

> a certain scepticism as to whether it will ever be possible to provide as rich a theory of cultural evolution as of biological evolution . . . Because of the complexity of human social behaviour, the rules of cultural transmission must inevitably be protean and difficult to define generally. This leads to a strong element of *ad-hoc*-ery in trying to interpret the facts.[40]

To a practitioner the extra complexity may be seen as a deepening of the basic approach, but unless this deepening produces germinal new discoveries – not simply incorporating findings from other disciplines in the style of Lumsden and Wilson – then one has what the philosopher of science Lakatos has called a 'degenerating research programme', namely an approach in which the increasing complexity of the theoretical assumptions fails to predict interesting new phenomena.

Human sociobiology may, or may not, turn out to be merely a scientific fad. It nevertheless presents grave political dangers because it offers the authority of science (once more) for a set of ideas that justify the inequalities of the status quo. Many popular books containing more or less sociobiological ideas have found a ready audience. Who today has not heard of Desmond Morris? Robert Ardrey and Lionel Tiger are not far behind, and major sociobiologists such as E. O. Wilson and Barash have themselves produced popular books.

Science is a very powerful and authoritative form of knowledge in our society. Arguments supported by appeal to this authority are consequently viewed as being powerful and difficult to challenge. Such popular works are therefore not just trivial, entertaining, irrelevant or absorbing. They are also providing a set of ideas within which political notions become justified: for instance, notions of the family as the 'natural unit' of solidarity and protection; or of the 'naturalness' of inherited wealth and property; or of kin solidarity ('our people') and its xenophobic counterpart; or of the traditional position of women in society, and of the essential 'natural' justice of inequality. Such ideas occur in the speeches of Thatcherite conservatives[41] and there even exist economic papers trying to demonstrate that monetarism is justified on sociobiological grounds.[42] In a more extreme form, these ideas underwrite National Front ideology. Richard Verrall, National Front theorist, claims that according to sociobiology:

> Altruism is directed primarily towards one's closest relatives, then to more distant kin and to large kinship groups such as nations and races. Abstract altruistic concepts directed towards 'mankind' and 'all races' are the products of our intellect, not our instinct.

From this he goes on to attack notions of the 'oneness of the human species' and to justify racism. He concludes:

> The great question of our time seems to be whether European man, the pinnacle of evolution, will destroy – through the unnatural [sic] notions which are the modern products of the intellect – what his inherited instincts have striven through aeons of time to preserve.[43]

Conclusion: an alternative view of human evolution

We have argued that human social behaviour and social organisation cannot adequately be explained by reduction to the action of genes, nor by looking for analogues in the behaviour of other animals, and we have shown that there are key aspects of human behavior that differentiate us from other animals. In our, and others' view, these key aspects form the bases for the novel evolutionary process that has transformed the world over the last few thousand years. This process is one, not of biological evolution, but of human technical and social organisation.

Humans, as we have seen, differ greatly from other animals in being biologically equipped both for conscious, rational, skilled solutions to problems, and for basing such solutions on shared experience through language. Why should humans have these characteristics? Because we are a species that satisfies its needs by an evolutionary novel process: work – that is, skilled social activity consciously planned for the satisfaction of future needs. There are many examples of biological adaptations that reflect our status as co-operating working animals; the areas of our brain devoted to language production and comprehension are examples. It is these special abilities that enable our creativity and our ability to bring about the great physical and social changes of our history.

Being human is about this capacity for social and environmental change. To be human is to be a flexible and adaptive animal. By focusing so much on what they see as the limits on human behaviour, sociobiologists deny, ignore and overlook much that is importantly human. The human being *is* special.

7. Animal behaviour to human nature: ethological concepts of dominance

BSSRS Sociobiology Group

The view that to understand society we must understand our 'animal nature' has a long history. Anyone reading an account in the popular press about, say, the behaviour of a football crowd, must be aware of how entrenched this idea is. Books of popular science, by authors such as Robert Ardrey and Desmond Morris, have encouraged this interest. This chapter looks at the political assumptions, and the political significance, of the way such writers treat one particular ethological concept, that of dominance and hierarchy. But first there is an assessment of how more academic works treat these concepts, showing that, far from the popularisers' gratuitously grafting reactionary politics onto good neutral ethology, many of the political assumptions are there in the primary works.

The concepts of dominance and hierarchy have come under criticism from ethologists. Nevertheless, they provide a valuable case study, since they are in practice still very widely used. Even the critical positions often retain political assumptions rejected here, and it is impossible to understand the shape of modern ethological work without realising the central position they have held in the development of the ethology of social relations.

Within animal groups there often seems to be a fairly stable pattern of threats and avoidances between particular individuals. The first serious study of animal social hierarchies was by Schelderup-Ebbe who, in 1922, described the 'pecking' order' of farmyard chickens. He found that when two birds are put together, one of them will quickly establish itself as the 'despot' (his word) in that it can peck the others with impunity. In a given flock it is possible to rank each member according to who pecks whom, the result tending towards (but rarely achieving) a linear chain of domination.[1] Within two decades of the publication of this paper, dominance hierarchies had been uncovered in most of the vertebrate families and the concept was later extended to invertebrates.[2] Even the easy-going chimpanzees of the Gombe Stream Reserve described by Jane Goodall in the mid-1960s have since been shown to display subtle but definite patterns of status interactions.[3] In consequence of the seeming ubiquity of the phenomenon, the study of animal social structure is now nearly synonymous with the delineation of dominance hierarchies. However, it is not their pervasiveness that interests ethologists as much as the correlations between an animal's position in the hierarchy and its access to valued resources such as mates and food. In this way dominance has been likened to territoriality in that 'both traits enable some individuals to obtain a larger than average share of available resources by disrupting competitors'.[4] Wilson defines dominance similarly: 'In the language of sociobiology, to dominate is to possess priority to the necessities of life and reproduction'. He has no doubts that 'this power actually raises the fitness of the animal possessing it'.[5]

In isolating dominance as something that can be individually *possessed* these authors are part of a long tradition dating back at least to the 1930s, which explains dominance hierarchies as the differential ownership of a unitary quality called dominance.

It must be said, however, that there has been a parallel tradition, almost as old, of debunking the idea of dominance as a unitary expression of genetic differences[6] and a number of researchers now prefer to think of status as something that animals actively strive towards rather than being passively driven by. The evidence of both traditions is examined later. It is enough for the moment to note that irrespective of whether dominance is treated as a genetic driving force or a social goal, material benefits are seen as accruing to animals because of and in proportion to their ranks rather than as a result of the particular relations they have with each other. Much intellectual effort continues to be put into finding out how rank *determines* social interaction and priority, with little consideration given to the possibility that the causal arrow points the other way.

Dominance then, is seen as a concise and simple way of describing agonistic social interactions (i.e. those related to fighting) between animals. We wish to criticise it from three different angles:

(a) Empirically, different measures of dominance often do not agree with each other;
(b) It overemphasises the aggressiveness of animal social intercourse;
(c) While it may be descriptively economical it is theoretically reductionist.

Dominance is inconsistent
Empirically, reducing the variety of competitive interactions seen in animal life to a single trait has been of patchy predictive help. Quite often, the correlations between rank and priority over resources are rather low. Male rhesus monkeys who win many aggressive encounters, for example, are not necessarily the animals who do most of the sexual mounting.[7] Moreover, correlations between different measures of dominance are often surprisingly poor.[8]

In addition the common assumption that rank itself is determined by individual attributes (such as sex, size, weight, hormone levels) has been disqualified by the observation that changes in some of these attributes occur more often than changes in rank.[9]

Dominance is divisive
One might have expected a developing science of animal social interaction to have concentrated, as a priority task, on explaining why animals form groups and how they co-operate. Yet as early as 1946, T.C. Schneirla complained that ethologists' enthusiasm for dominance and aggression encouraged 'a particularistic static type of thinking about social organisation actually distracting from the essential problem of group unity'.[10] Twenty-five years on, J. H. Crook finds it 'curious' that in spite of an emphasis in some earlier writings on mutual aid in animal societies, 'social ethology

has not flowered in the clear-cut manner of physiological ethology'. The reason, he thinks, 'lies in the failure of ethologists generally to consider social behaviour as a group process but instead in terms of stereotyped/reciprocal interactions between individuals'.[11] Neil Chalmers suggests that the trouble with dominance is that 'people have tended to see it everywhere. Whenever two adult . . . animals within a group have gathered together, their interactions have been interpreted, as often as not, as manifestations of dominance . . . We need to counter this tendency.'[12] The dominance concept has had the effect of press-ganging complex, rich and co-operative interactions in the service of a simple linear ordering based on competition.

Dominance is reductionist

Whatever else dominance might be, workers in the area agree that it is a social relationship between individuals. One cannot infer the dominance of an animal isolated from its companions: it is a relative rather than an absolute concept. This simple fact wholly undermines theories that posit dominance as a genetic trait available for selection by evolution; for how can an individual inherit a relationship? Bernstein puts the point succinctly:

> Since dominance must be determined in a social context it cannot be abstracted as an attribute of an individual. If dominance were an attribute of an individual, and if selection strongly favoured dominant animals, then would it not be reasonable to expect that the proportion of dominant individuals would increase? Can one have a population consisting only of dominant individuals and still measure relative relationships? Agonistic dominance relationships cannot be selected for because they cannot be divorced from their context.[13]

An example of the kind of theoretical contradictions that can arise when dominance as a social relation gets tied to genetic explanations in terms of reproductive fitness is evident in some statements on dominance relations between the sexes, maintaining both that dominant animals are fitter than the animals they dominate, and that males are dominant to females. Thus, 'in the great majority of primate species, males appear more aggressive than females . . . and are generally dominant to them'.[14] By classic sociobiological reasoning we should expect that male primates will have a selective advantage over females and that 'genes for maleness' will spread through the population – an evolutionary nonsense for a sexually reproducing species. Obviously, the authors intend to imply nothing of the sort. They probably meant to say that among primates, it pays to be relatively aggressive *if one is male*. The contradiction arises because the authors have not properly understood that dominance is a social relation between individuals rather than a property they possess. Labelling one partner in a social interaction as dominant need imply nothing about its fitness relative to the other. The social relationship between males and females needs to be understood as a compromise (or if you prefer, a mixed evolutionary stable strategy) between what is to the evolutionary advantage of both.

There has, however, been a drift away from theories that locate dominance in individual physiology and a welcome re-emphasis on explaining dominance in specifically social terms. Thelma Rowell, for example, criticises those ethologists who set up artificial competitions between animals as a quick way to measure their supposed rank. By throwing scraps of food between animals, she points out, 'a hierarchy may be caused rather than revealed'.[15] In other words, it is not dominance that leads to the observed antagonistic interactions, but observing competitive interactions that leads ethologists to infer dominance. Dominance also depends on the social context. When two rhesus males that were the highest ranking in their group were removed and introduced to a new group, they were dominated by all the males in that group, but when returned to their home group, the original relationships were re-established. Similarly, the highest ranking males of the second group behaved submissively to all males when they were moved to the first group *even though they had previously dominated two of its mem-*

bers. Monkeys who have been used to dominating particular individuals in one social context may submit to them in another.[16]

The trend away from genetic determinism has focused attention on the social texture of dominance, the complex way it develops between animals and the way these relationships link together in a stable social structure. In rejecting the idea that animals are the puppets of their genes, it is however becoming common to embrace the 'sociological' attitude that animals are conscious of each others' ranks and constantly on the look-out for promotion up the social ladder. This brand of anthropomorphism has marked the introductory textbooks for some years (see examples in table 1.1) but it is also to be found in scientific journals. 'Social status' according to T. D. Wade 'will in fact be probably best understood as dynamic goal-directed process'. Describing an alliance between two monkeys he says its 'primary aim . . . was apparently to raise rank'.[17] 'Similarly, a variety of evidence indicates that monkeys, like their human observers construct rank hierarchies of the members of their group'.[18] Also, 'adult females and immature chimpanzees enjoy special rights or favours granted by the adult males' and 'it is not always the actual possession of the item that gives rise to fighting, but rather the infringement of pretended rights to the item'.[19]

It is reasonable to hypothesise that animals seek social as well as material rewards, and that they may in some sense be conscious of their social relations. *But it does not automatically follow that animals see their social relations as relations between their ordinal ranks*. Embedded in the notion of status-as-a-goal is the conviction that animals are aware of their respective positions along some quantitative scale ranging from low to high. With very little trouble the ousted unitary genetic quality has ensconced itself as an apparently quite acceptable unitary social quality; the quantitative, reductionist nature of dominance theory stays the same. The argument may be correct that monkeys preferentially seek the company of other monkeys which the observer considers to be high ranking, but we have no reason to suppose that it is 'rank' as such that is attractive, or

Table 1. Anthropomorphism in textbooks

J. D. Carthy (1966) *The Study of Behaviour*	'there is a dominant individual who has certain rights because of his rank' (p. 42)
D. E. Davis (1966) *Integral Animal Behaviour*	'animals know their rank and stay in their place' (p. 69 and also p. 72)
P. Dolhinow (1972) *Primate Patterns*	'the dominant male [has] the right both to fight and to deny the right to fight to any other member of the group' (p. 379)
I. Eibl-Eibesfeldt (1975) *Ethology, the Biology of Behaviour*	'a social hierarchy presumes . . . that some members will seek authority' (p. 390)
W. Etkin (1969) *Social Behaviour and Organization among Vertebrates*	'the precedence involved is not one which is fought over, but is a regular characteristic . mutually agreed upon' (p. 14 and also p. 13)
A. Jolly (1972) *The Evolution of Primate Behaviour*	'each animal knows the other's strength and respects his rights' (p. 172)
R. A. Maier and B. M. Maier (1973) *Comparative Psychology*	'one animal is dominant . . . and is allowed a number of privileges' (p. 60)
A. Manning (1972) *Introduction to Animal Behaviour*	'high rank . . . determines access to food' (p. 260) and 'young males challenge for positions of dominance' (p. 265)
A. Portmann (1961) *Animals as Social Beings*	'animals themselves recognise their different "degrees" and deal with each other accordingly' (p. 81)
J. C. Ruwet (1972) *Introduction to Ethology*	'the alpha enforces respect for its right by force' (p. 175)
J. P. Scott (1972) *Animal Behaviour*	'young males learn their place in the dominance order' (p. 185)
E. O. Wilson (1975) *Sociobiology: The New Synthesis*	'equally ranked subordinates contend for the top position' (p. 287)
J.C. Wynne-Edwards (1962) *Animal Dispersion in Relation to Social Behaviour*	'the top bird normally has the right to peck all the other birds without being pecked in return' (p. 134)

that monkeys have a model of rank at all. This may seem to be mere carping on the form of words used. But wording has important consequences for the way we think about animal social structure and the way that popular writers portray animal life to lay audiences. With words like 'rights' and 'respect' bandied about the literature, it is no wonder that popularisers have painted animal social life as the bourgeois world writ small. The ethologists have already primed the canvas.

We have reviewed two kinds of explanation of dominance hierarchies – in terms of individual genes and of individual aspirations. They may seem antagonistic but they share a common philosophy. The richness of social structure is reduced to a few thin abstractions which do not rise above the level of the individual. This is about as helpful as equating consciousness with an inventory of properties of the neurones in the brain or arguing that each neurone is individually conscious. The work of Bernstein and others, however, has shown that the properties of an animal social group are more than the sum of its parts. What ethologists call 'dominance' is their interpretation of a pattern running through animal communities whose underlying components vary widely. What counts as dominant behaviour in a flock of hens will be inappropriate among a flock of sheep. In human society, where social organisation is vastly more intricate and dynamic, explanations in terms of social structure have even greater power over discussions which deal only in individual components. Yet in humans also, theorists have treated attributes of social structure as if they belonged to private individuals. Economic inequality, for instance, is often excused as the outcome of inherent individual differences rather than as definite social relations between classes of people actively maintaining a particular society at a particular time in its history.

Probably the most plausible reason why status and hierarchies are seen to play so important a role in animal societies, is that they loom large in our own society. A kind of anthropomorphism has occurred. Now, scientists must get their models of nature from somewhere, so it is understandable that they might borrow ideas close to hand in the society around them. To a great extent therefore anthropomorphism is unavoidable.[20] But all too often, ethological anthropomorphism occurs quite unconsciously with the result that status and hierarchy seem to be truly natural features of the animal world, instead of more or less adequate intellectual manufactures. Along come the popularisers who then extrapolate from the ethological literature back to human society and conclude that we are all naturally hierarchical and status oriented.

Let us now look at how popularisers of animal behaviour draw out the political implications of the ethologists' seemingly innocent thoughts about dominance.

Dominance popularised

A rash of popular books appeared during the 1960s and 1970s which attempted to prove the animal nature of human nature by introducing the general public to the social habits of monkeys and apes. Some common themes run through them all: (a) they stress that our social preoccupations are genetically determined and inescapable; (b) they completely ignore recorded human history and concentrate wholly on an evolutionary time-scale; and (c) the animals they write about conveniently seem to share the authors' own political beliefs. We can consider each of these in turn.

Genetic determinism

The popularisers seem determined to convince their readership that dominance is innate, that there is nothing we can do about it. 'We know that the true alpha is born not made', says Robert Ardrey in *The Social Contract*. Actually, we know nothing of the kind. In the first place, no one has ever mounted a study on the heritability of dominance (as noted by Rowell);[21] and besides, if evidence existed of genetic variation between individuals accorded different ranks, it would not prove that dominance is an attribute of the genes. The popularisers contend that 'dominance is a natural instinct' by merely noting that animals related to our ancestors behave 'dominantly'. Thus in Ardrey's *African Genesis*, 'hierarchy is an institution among all animals and the drive to dominate one's fellows is an instinct 3 or 4 million years old'.

We are also told that dominance is something which individuals own in different amounts. Lionel Tiger and Robin Fox, in *The Imperial Animal*,[22] speak of the non-uniform distribution of capacities for dominance behaviour which they compare to IQ, while Ardrey tells a story about a rhesus male 'of almost unbelievable dominance [whose] factor of dominance compared to number last [the lowest ranking animal] was about 50 . . . a normal maximum would be about 5, (*African Genesis*).

History is ignored

All of these books choose to ignore written records of our past. They concern themselves wholly with pre-history about which we know comparatively little. They are contemptuous of real differences in content, preferring to concentrate on superficial similarities of form. Thus a baboon hierarchy = a feudal = a capitalist = a 'communist' hierarchy. Hierarchy becomes a synonym for all social structures. These authors typically make direct extrapolations from animals to people as though the space between were a fiction dreamed up by historians.

Thus, the hierarchical system is 'the basic way of private life' and 'behind the façade of modern city life there is still the same old naked ape—only the names have changed' (Desmond Morris, *The Naked Ape*). Tiger and Fox insist that, 'It must be understood that the process that gives rise to empire is the *very same* process that primates engage in simply in order to exist and persist.' In the zealous hunt for similarities glaring differences are blindly passed over. It will take more than Ardrey's repeated declarations that the USSR, the USA and baboon troops all have hierarchies, to conclude that they are all the same underneath. Power for instance is won in one system by owning capital, in another by toeing the party line, and in another by physical intimidation. In any case, to the extent that the USSR, the USA and baboons do share common features, there does not have to be a common biological explanation. A convincing explanation of the resemblances between the USA and USSR must at least consider the similar problems they face, the similar historical experience they have to draw on, and the common resources they are competing for.

Political anthropomorphism

Animals are seen as engaging in all sorts of behaviours that are regarded as typical of human beings – jostling for status, feeling ambitious, and so on. John Bleibtreu is explicit about the parallels:

> Social recognition comes to animal leaders in much the same way as it does to human leaders, by the community's consent to the leader's competence. Within each stable animal society there exists a scale of interlocking social relationships and since the acquisition of leadership (or dominance) requires that these relationships be manipulated, animal leaders are equally as political as human leaders.[23]

Granted, animals do have stable relationships which we interpret as hierarchies. Field studies leave no doubt either, that in many primate groups a flux of alliances is constantly being made and dissolved between pairs of animals in which one partner helps the other to defeat a third. From the human onlookers' point of view it may look very political but we do not know that animals see it the same way. To be political requires an awareness of social relations in the abstract, divorced from the actors who embody them. Just because animals follow a given individual, it does not mean they have a concept of a 'leader' or that they formally consent to be led. Just because an animal may aspire to more congenial, advantageous relationships with its fellows, it does not mean it craves status, dominance, power or any other unidimensional abstraction. Nor does it follow from the fact that friends are useful, that influence is *cultivated*, or from the fact that aid is often reciprocated between friends, that one good turn *deserves* another.

It should be clearer now why so much space was given above to bemoaning the way ethologists have given the impression that animals are aware of their position in a hierarchy. In the textbooks and popular books this theme is developed to the point where animals appear to recognise civil liberties. Animals accord themselves rights it seems, and they receive the privileges due to their respective social ranks. The literature is cluttered with these sort of statements, some of which are given in table 1.2. Even the Chambers' dictionary defines a pecking order as a 'social order among poultry according to which any bird may peck a less important bird but must submit to being pecked by a more important one'. It sounds as if chickens have meetings to agree on the order of importance and to dish out the honours.

This is significant not only because it misleads us in our attempts to understand animal behaviour. It acts as a 'justification' of our own forms of social organisation by making them appear natural, an inbuilt part of our biological make up, which we hold in common with other animals.

The popular authors have very definite political preferences. They get quite out of breath enthusing about the essential fairness of our society compared to feudalism. By all accounts, class barriers have been broken down opening the way for promotion of any individual who has the talent and the will to succeed, while meritocracy is (and always has been) the ultimate evolutionary goal. Predictably, animal society led the way. 'Equality of opportunity is the second law of vertebrate society,' announces Ardrey on the first page of *The Social Contract*. The crunch comes later when we are told that whether you are monkey or man [sic], if you do not earn enough, it's nobody's fault but your own: 'boy or baboon granted equal opportunity to display its worth must be willing to settle for less' (that is, to accept lower rewards than others).

According to Knipe and Maclay, money is simply a way to measure out our different degrees of rank: 'There is a fundamental relationship between wealth and dominance.'[24] Tiger and Fox tell us that, 'money is an important indicator of personal worth, particularly in the basic sense that one can earn it or cannot. Welfare money is not the same as worked-for money.'[25] One would suppose from reading John Mackinnon's *The Ape Within Us*,[26] that it is only an accident of birth that prevents apes from independently evolving a system of commodity exchange:

> There is nothing in the behaviour of wild apes that parallels barter or commerce . . . Captive apes do, however, have a keen sense of comparative values and I am sure that it would be quite easy to encourage the great apes to reorganise their lifestyles on a commercial basis.

Table 2. The concept of dominance in popular books

J. Berrill (1970) *Wonders of the Monkey World*	'the dominant male leader has undisputed right to first choice of food' (p. 39)
G. H. Bourne (1972) *The Ape People*	'Onan decided to stake his claim for the number one position' (p. 208, also p. 209)
J. Goodall (1971) *In the Shadow of Man*	'each individual knows his place in the social structure' (pp. 111, 114, 116)
G. B. Schaller (1968) *The Year of the Gorilla*	'every animal knows clearly where it stands in relation to every other animal' (p. 148)
L. Williams (1967) *Man and Monkey*	'Jojo's allegiance to the rules of dominance will always come first . . . [there is] great respect for the dignity of his position' (p. 93, also p. 95)
R. Ardrey, (1961) *African Genesis*	'rank must come first in the preoccupation of any social animal, for rank tells all . . . all is determined by the single acquisition of status' (p. 96, also pp. 11, 91, 97, 98, 99)
(1967) *The Territorial Imperative*	'alpha may peck beta, but beta may not peck back' (p. 223)
(1970) *The Social Contract*	'male animals . . . compete for symbols like . . . high rank in hierarchy' (p. 194, also pp. 107, 108, 117, 131, 153, 155, 170, 106)
R. Dawkins (1976) *The Selfish Gene*	'peck order — a rank ordering of society in which everybody knows his place and does not get ideas above his station' (p. 122)
A. Jay (1972) *Corporation Man*	'animals know their status and the status of all the others in the group' (p. 140)
H. Knipe and E. Maclay (1972) *The Dominant Man*	'every individual knows its proper place' (p. 4, also p. 6)
K. Lorenz (1967) *On Aggression*	'all social animals are "status seekers" (p. 36)
E. Morgan (1972) *The Descent of Woman*	'when a baboon is manoeuvring his way up the rank order in the hope of becoming dominant' (p. 232, also pp. 211, 214)
D. Morris (1967) *The Naked Ape*	'fighting is used . . . to sort out dominance disputes' (p. 126, also pp. 146, 175)
(1969) *The Human Zoo*	'other species also indulge in intense status struggles and the attaining of dominance is often a time-consuming element of their social lives' (p. 77, also pp. 41, 42, 62, 64)

Dominance is thus turned into a commodity. Knipe and Maclay hypothesise that, 'if all consumable goods and items of economic value can be *bought directly with status points*, then every member of the community can arrange to *spend his dominance* when and how he chooses' [emphasis added].[27]

The real business of how wealth is generated by labour under capitalism – the social relations of exploitation between owners of capital and their employees – is wilfully evaded. We learn only that rich people are rich because they are dominant, and they are dominant because they work hard. We are not enlightened as to why 12 hours on the night shift is not equivalent to a few minutes of high finance. Tiger and Fox tell us that, 'true democratic theory advocates . . . that the basis for striving should not be arbitrary, that the basis for being a competitor should not be fixed at birth by class, sex or colour; that people should be allowed to dominate each other on their merits, not on factors beyond their control.'[28] True democracy, they say, is about mutual domination, not co-operation.

Human hierarchies

As an antidote to the popularisers' relentless allusions to the universal sameness of social architecture, we shall stress here its extraordinary variety among human communities and the profound qualitative differences between animal and human social being.

Anthropologists are continually impressed by the diversity of social arrangements which have evolved in human civilisations around the world. Compare our society with, for example, the Hadza of Tanzania: they live in groups varying in size up to a hundred members. A group has no acknowledged leader and its members do not constitute a stable unit. A group will remain together for a few weeks at most, before some members move off to a new site or join other groups. The Hadza are not unique. Similarly fluid social organisation is found in other hunter–gatherer societies and there is often little or no hierarchical structure.

Even accepting that many societies are organised in a very loose sense hierarchically, how does evolutionary theory begin to account for historical changes in the form that hierarchies take – such as, changes from chieftainates to slave societies, classical Chinese societies, feudal and caste systems, capitalist and communist states?

The popularisers ignore, or do not understand, that only in human societies, do members recognise *hierarchy* as existing on a level that is not reducible to personal relations between individuals. Hierarchies involve relations between *positions* in the social structure which exist independently of the particular individuals. The economic relationship between serfs and their feudal lords, for example, cannot be said to have depended on their respective personal qualities. The power structure of modern Western capitalist society is qualitatively different from feudal society and neither can be said to be mere variants of a basic human societal type. Every form of human social organisation has specific historical roots. Historian Perry Anderson, for example, traces the origins of Western capitalism to the merchant and guild craft classes of European feudal society – itself a specific historical entity deriving from Germanic tribal structure and the legacy of the Roman empire.[29]

A common characteristic of many forms of human society that the popularisers do not notice is that the majority of the population work not only to support themselves and their families but also to keep an additional class of non-producers who have power over them. This is clearly seen in feudal and slave societies. It is equally true, but less transparently so, of our own capitalist society and in Eastern communist states. We need not argue, however, that it is an inherited biological part of the human condition since it is not evident in 'primitive' hunter–gatherer societies. What other explanations are there for the evolution of unequal distribution of rank, goods and work?

It has been argued that it is both a cause and a consequence of the intensification of production which succeeded the relatively leisurely, individualist and generous life-style of the hunter–gatherers.[30] At a transitional stage (where societies have ruling chiefs) production begins to be more communally organised and able to create a surplus. But the organisers – the chiefs – remain tied to the membership by a complex network of mutual gifts, and the position of chief is assumed by a very different process from, say, a feudal lord. Clearly, then, in human societies, rank, power and privilege are not biologically pre-destined; they are politically contingent.

We have seen that ethology is all too often used to support the idea that the inequalities of our current social order are inevitable and natural, and the belief that however hard we try to eradicate them, 'human nature' will reinstate them. The spread of such ideas can only undermine the belief that by our collective co-operative struggle we can restructure our society.

8. Population, poverty, and politics

Richard Clarke

Population biology is an important area of ecology, and forms a major component of ecology textbooks[1] and university ecology courses. It impinges upon almost every other area of ecological theory, linking closely with ecological energetics and production ecology on the one hand, and with evolutionary and behavioural ecology on the other.[2] Moreover, it articulates the 'central dogma' of modern ecological theory, located by the twin axes of resource scarcity and 'limiting factors'. Today the notion of scarcity has come to dominate popular thinking and provides an 'ecological' interpretation of the environmental crisis (and even, according to some ecologists, the economic and social crisis as well). Central to such 'socio-ecology' is the way that the weight of human numbers is so often given as the central, or at least the major, cause of social ills ranging from environmental pollution and resource scarcity to poverty, malnutrition, and the growing gap between 'rich' and 'poor' nations.

Population ecology is an example not only of the use of ecological ideas to explain social phenomena; much ecological theory itself can be shown to have been influenced by social ideas, and in some cases to derive directly from importing social theories into ecology. Although the context of population ecology is often presented primarily as a matter of mathematical and field technique, the wider (ecological and social) context in which these techniques are situated illustrates the process by which biological theory is both shaped by social and political values, and also in turn influences, and is used to justify, social ideas and policies.[3]

Malthus

Much current thinking in both ecology and social theory can be traced back almost two centuries to the work of Thomas Malthus which established the twin notions of limited environmental capacity (carrying capacity) and the struggle for existence (see chapter 9). Malthus' argument in his first (1798) *Essay on the Principle of Population*[4] was in essence that the fundamental and inevitable checks to population growth were limited resources. The existence of poverty and inequality was therefore inevitable, and social reforms aimed at a more humane and just society were doomed to failure. More specifically Malthus argued against poor relief and other measures to alleviate suffering amongst the poor, aged, or sick. He argued that such measures merely perpetuated poverty – by permitting the poor to survive and breed!

Malthus' work was by no means original but he provided the first popular elaboration of a coherent theory of demographic change and its political economy, at a particularly propitious time in England. It was only after Malthus that the notions of 'population pressure' and 'environmental resistance' (to use the modern jargon) became subjects of general discussion and gave rise to scientific enquiry.

Two points are worth emphasising here. The first is that Malthus, anticipating the practice of sociobiology today (see chapter 6), appealed to nature as the legitimator of social structures, relationships and of the status quo by making a direct comparison between the behaviour of animal and plant populations, and those of humans. After enunciating his 'principle' of population:

> The power of population is infinitely greater than the power of the Earth to produce subsistence for Man . . . population when unchecked, increases in a geometrical ration. Subsistence only increases in an arithmetical ratio.

He declared:

> The race of plants and the race of animals shrink

> under this great restrictive law. And the race of Man cannot, by any efforts of reason, escape from it. Amongst plants and animals, its effects are waste of seed, sickness and premature death. Among Mankind – misery and vice.[5]

The second important point is that Malthus' writings were (unlike those of many later Malthusians) quite explicitly political tracts. Malthus declared that the poor and labourers themselves were to blame for their situation:

> The principal and most direct cause of poverty has little or no direct relation to forms of government or the unequal division of property and . . . as the rich do not in reality possess the power of finding employment and maintenance for the poor, the poor cannot, in the nature of things, possess the right to demand them.

He went on to confess that the purpose of his writing was the hope that:

> every man of the lower classes of society, who became acquainted with these truths, would be disposed to bear the distresses in which he might be involved with more patience, would feel less discontent and irritation at the government and the higher classes of society on account of his poverty, and would be on all occasions less disposed to insubordination and turbulence.[6]

It is unlikely that many of the 'lower classes' were convinced by Malthus, but his work nevertheless provided the ideological justification for reactionary social theories and repressive policies. The most infamous example is the Poor Law (Amendment) Act of 1834. This established the workhouse system (in place of the earlier and relatively benign system of out-relief) as the only ultimate recourse for the poor, the sick and the infirm. The workhouse regime was designed so that every able-bodied inmate was 'subject to such courses of labour and discipline as will repel the indolent and vicious'.[7] It was designed as a cure for poverty by relieving the state and property owners of its consequences, by making those consequences horrific in the extreme and by discouraging – and in the workhouse system, preventing – breeding.

This first major political impact of Malthusian theories came in the 1820s and 1830s – at the time of the first major recession in industrial capitalism. Certainly it is significant that the latest resurgence of 'scientific' justifications for class, race, and sexual inequalities and oppression seem to accompany a worsening of the crisis of the society which they seek to justify.

Malthus' theory was not only influential politically. It also established a 'research programme' in biology which has had a central influence – particularly in population biology – ever since. It was certainly central to the development of evolutionary theory and was acknowledged as a major influence by both Darwin and Wallace[8] (see chapter 9).

Population theory in the twentieth century

The recent history of population ecology dates from the 1920s and has two related aspects. The first involved attempts by ecologists to investigate the factors controlling animal populations in the field. A major stimulus here was the need to provide a theoretical basis for range management, forestry, and insect pest control.

In parallel there came attempts to develop mathematical models of population growth and regulation.[9] These however, arose from studies not of animal but of human populations by Pearl and Reed. They popularised the logistic (sigmoid or S-shaped) curve and its mathematics which are fundamental to modern population ecology and will be familiar to all students of this subject. Pearl and Reed's work had actually been anticipated by that of Verhulst almost a century before. In 1845, taking up the suggestions of his mentor Quetelet a decade previously, Verhulst developed a mathematical model of the Malthusian limitation of human population growth due to the constraints of crucial resources (in this case farmland) in which an exponential rate of growth was retarded by a linear function proportional to the size of the 'excess' population.

Several points are of interest here. The first is that whatever the correspondence of the curve to the real populations to which he tried to fit it, its mathematics do not correspond to the demographic argument which he advanced in its support – the Malthusian argument that a population would increase exponentially until such time as crucial resources became limiting, which implies that growth shifts abruptly from an exponential rate to a slower 'logistic' one at some crucial density. Second, Verhulst himself realised that there were alternatives to the 'logistic' equation which was only one of a number of mathematical possibilities. For example, 'environmental resistance' could be proportional to the square of the 'excess' population, or to the ratio of the 'excess' to the total population. Quetelet himself had suggested that environmental resistance was proportional to the square of the speed with which the population increased; this appealed to him because the relationship was analogous to the resistance that a medium (e.g. water) opposes to a body travelling through it.

Raymond Pearl's independent 'rediscovery' and popularisation of the logistic curve in the 1920s arose out of his wartime work in Hoover's Food Administration programme. In collaboration with Reed he published in 1920 an article[10] applying the sigmoid curve to US census data and subsequently, when he learnt of Verhulst's work, adopted the term 'logistic'. Unlike Verhulst, this 1920 paper did not derive the curve from Malthusian assumptions, but in keeping with the precepts of 'scientificity' proposed the curve to fit their data and then tacked on their Malthusian assumptions as an 'explanation'. Nevertheless the implications were clear – given a limited area into which a population could expand, the rate of increase at any time was proportional to the magnitude of the population and to the 'still unutilised potentialities of support existing in the limited area.' As these 'potentialities' were reduced, the rate of population growth would fall. That the observed reduction in US birth rate and family size could be due to *social and historical* factors rather than Malthusian limits was clearly a possibility that they preferred not to consider.

Moreover Pearl and Reed did not regard their 'logistic' curve as an empirical fit of mathematics to data, but rather as a *law* of population growth which could be used predictively. In this, Pearl concurred with his mentors, the (appropriately named) Herbert Spencer Jennings, and the biometrician and eugenicist Karl Pearson, with whom he had studied. All saw science as consisting of the classification of 'facts' and the deduction of 'laws' which involved modelling the relationships between them. Such a view of science as an objective and 'neutral' (and essentially mathematical) activity was, of course, useful to a profession which had become increasingly involved in the mathematicisation of social Darwinism as a new 'scientific' racism and justification of class society.

The implication of Pearl's work was that there were real, qualitative, and not merely mathematical or quantitative, similarities in the growth curves of bacterial populations in a test-tube and human populations in society. Not surprisingly he was extensively criticised on precisely this point. It was pointed out that many human demographic histories simply could not be fitted to a logistic curve. And even where they could, the reduction in growth rate could more accurately be attributed to birth control or to social, ethical or economic changes rather than to the effects of population density and resource limitation. In 1931, Lancelot Hogben pointed out that 'the mere fact that the same type of equation can be used for two different sets of variables does not necessarily denote the same intrinsic mechanism'.[11] Interestingly, Pearl's own work was based on the earlier work of a physiologist T. B. Robertson who had published articles in 1908 applying the sigmoid curve to animal, plant and human populations, and of whom Pearl had himself been highly critical on precisely this point. The reversal of his position indicates how strongly Pearl had become attached to his logistic hypothesis.

However, its subsequent general acceptance was not merely due to Pearl's own vigorous promotion[12] but to explicitly political patronage. From 1925 to 1930 Pearl was the director of the Institute of Biological Research at John Hopkins University, a unit created with a grant from the Rockefeller Foundation essentially in order to allow Pearl to pursue his research with complete freedom.

Pearl's work formed the basis for the mathematical theories of competition in association with the names of Lotka,[13] Volterra, and Gause. These developments in theory were accompanied by a great deal of work on isolated laboratory populations in highly artificial conditions, such as that of Chapman on flour beetles, and Gause and others on protozoa, yeasts and mites – all aimed primarily at establishing the applicability of the Verhulst–Pearl equations and their derivatives. These simplistic mathematical theories and experiments subsequently became in the 1960s the foundations for a reductionist paradigm in ecology based upon scarcity and competition (see chapter 9).

In parallel with these mathematical developments came attempts to understand and to develop theoretical models to explain the way in which animal populations are regulated in the field. Early debates raged around the mechanisms of population regulation in natural populations. Some workers such as Uvarov[14] emphasised the instability of field populations and ascribed this to the dominant role of physical factors such as climate. Others followed A. J. Nicholson[15] who emphasised the 'balance' of animal populations. Using Pearl's 'logistic curve' he argued that a species' population size and growth rate were related to the carrying capacity of the environment and this required the primacy of *density dependent* agencies, the most important of which was competition, as the crucial regulatory factor. These debates continued well into the 1960s, the different 'schools' being represented by workers such as entomologists Andrewartha and Birch[16] who argued that populations were generally controlled by 'catastrophic' factors such as weather, and the ornithologist David Lack[17] who championed the importance of density-dependent mortality due to food shortage in the early stages of life.

Part of the interest in such debates lies in their co-existence as competing 'paradigms' in ecology, a co-existence which can be explained in part by the organisms and habitats with which their authors were dealing. Recently it has become evident that different types of regulatory effects predominate in different types of environment.

The metaphors magnify

These developing views on animal population regulation inevitably became linked to concern about rapid human population growth.

Since the days of Malthus and before, natural catastrophies have been used to 'explain' disease and starvation as 'natural' checks on human numbers. But the recognition that density-dependent factors are the most important *regulatory* mechanisms in natural populations has been used to argue that an increase in density-dependent mortality, as some supposed fixed 'carrying capacity' is approached, is the parallel 'explanation' for poverty and hunger in human societies.

Lack illustrated his principle of density-dependent population regulation with the 'example' of human living standards. In the UK, he declared, the 1950–51 National Food Survey showed that amongst working-class people earning less than £10 per week, the amount spent on food per head decreased from £1.20 per week for a childless couple to £0.60 per week for a family with three children. Meat consumption similarly decreased from 34oz to 17oz per head in each case. In Australia the percentage of families in 1957 with a 'first-class diet' was 73 per cent where there were two children and only 16 per cent where there were six children.[18] Lack's clear implication is for the operation of a Malthusian check, but there is absolutely nothing in his data to suggest that poverty in large families was the result of having many mouths to feed.

Ideology, ecology, and the politics of population control

To deny the existence (let alone the possibility) of the need for a reduction in the world rate of human population growth is as futile as to deny the need to abolish the disparities between rich and poor – both between nations and within them. But as in the latter case so in the former; what is needed to change the situation is an understanding of it – and understanding is what is so patently lacking in popular debates on the 'population explosion'.

Today's resurgence of crude neo-Malthusianism is characterised by the opening paragraph of Paul Ehrlich's *Population Resources, Environment* (subtitled *Issues in Human Ecology*[19] which enjoyed a considerable vogue in the early 1970s and is still found today as a standard text for college courses on the environment:

> The explosive growth of the human population is the most significant terrestrial event of the past million millenia. Three and one-half billion people now inhabit the Earth, and every year this number increases by 70 million. Armed with weapons as diverse as thermonuclear bombs and DDT this mass of humanity now threatens to destroy most of the life on the planet. Mankind itself may stand on the brink of extinction; in its death throes it could take with it most of the other passengers of Spaceship Earth. No geological event in a billion years – not the emergence of mighty mountain ranges, nor the submergence of entire subcontinents, nor the occurrence of periodical glacial ages – has posed a threat to terrestrial life comparable to that of human overpopulation.

Ehrlich is merely one of the more popular and prominent prophets of eco-catastrophe who put population growth firmly at the top of their list of major world problems. His book *The Population Bomb*[20] which sold several hundred thousand copies following its publication in 1970, declared that 'the battle to feed humanity is over'. Similarly, prominent media ecologist David Bellamy declares:

> The only way ahead is through population control. If we don't control the population, or contain the population resource, then we are fiddling while Rome is burning.[21]

The tradition is a long one going well back to Malthus. The post-1945 resurgence of neo-Malthusianism may be dated from Paul Vogt's *Road to Survival*[22] published in 1948. Vogt argued that unless the 'untrammelled copulation' of 'spawning millions' was brought to an end, 'we might as well give up the struggle'. Following Malthus in his 'ecological' determinism, Vogt also made his political position quite clear, declaring that it was necessary to get rid of the 'sort of thinking . . . that leads to the writing and acceptance of documents like the *Communist Manifesto* . . . It tricks Man into seeking political and/or economic solutions.' It was no use trying to build a better world, he argued, because our environment is 'as completely subjected to physical laws as is a ball we let drop from our hand'.

In the 1960s, with the growth of the environmental movement, such analyses began to take on an 'ecological' flavour. The ideological implications (and practical applications) of this new 'socio-ecology' are well illustrated by the writings of Garret Hardin (a popular spokesperson for the US 'ecology' movement) in his well-known paper, 'The tragedy of the commons'.[23] Hardin compares the environment and the Earth's resources to a pasture, open to all. It is 'natural' says Hardin, that each commoner will want to keep as large a herd as possible, which causes no problems for centuries 'as long as tribal wars, poaching, and disease, keep the numbers of both man and beast well below the carrying capacity of the land'. Ultimately, however, 'comes the day of reckoning . . . when the long-desired goal of social stability becomes a reality. At this point, the inherent logic of the commons remorselessly generates tragedy.'

As rational beings, each commoner examines the advantages of adding just one more animal to the herd. Since milk, meat and progeny accrue solely to the individual who, however, suffers only a fraction of the disadvantages resulting from overgrazing which are shared by all the commoners, the only sensible course of action is to increase the herd to the maximum. But every other commoner is in the same position. 'Therein lies the tragedy,' declares Hardin, 'each man is locked into a system that compels him to increase his herd without limit – in a world that is limited.' The pasture eventually collapses under the strain and can support nothing.

Such an 'ecological' analogy conveniently explains away the problems of environmental pollution and resource depletion and conceals the fact that we do not have a world of 'commoners' with equal access to resources, but an economic system in which the commons have been stolen from the people by large multinational companies and are being exploited to produce profit rather than to satisfy human need. Of course, the parable *could* be taken as an argument for socialist management of the commons, in the interests of the commoners as a whole. But it leads Hardin to declare that in a commons, 'freedom to breed is intolerable' and to call for 'mutual coercion, mutually agreed upon'. And, later in the essay, Hardin declares, 'conscience is self-eliminating' and, using arguments that could have been taken straight from some eugenecist pamphlet of the late nineteenth century or of Hitler's Germany or of the National Front today, continues in true social-Darwinian fashion:

> People vary. Confronted with appeals to limit breeding, some people will undoubtedly respond to the plea more than others. Those who have more children will produce a larger fraction of the next generation than those with more susceptible consciences. The differences will be accentuated, generation by generation.

For Hardin, the message is simple. Coercion, compulsion is the only way, and this must be directed against the 'irresponsible'. And in case we should have any qualms:

> Coercion is a dirty word to most liberals but it need not forever be so. As with the other four-letter words, its dirtiness can be cleansed away by exposure to the light, by saying it over and over without apology or embarassment.[24]

By *whom* and against *whom* should such coercion be applied? The answer is, of course, by those who have privileges and the power to protect them. As Barry Commoner says in the *Closing Circle*:

> One does not have to dig very far back in the literature of the population control movement . . . to discover alarming indications of racial and genetic discriminations; and the social and political implications for what some population control campaigners now say are frequently alarming.[25]

The extreme consequences of Hardin's ideas were to be found in Hitler's policies on 'Lebensraum' and towards the Jewish peoples – or today in South Africa where one-fifth of the population is encouraged to procreate 'for the good of the nation' and four-fifths discouraged from doing so, also for the 'good of the nation'.

Such ideas are also being actively promoted today in respect of the relationship between 'rich' and 'poor' nations of the world. In their best-selling *Famine* 1975 the Paddock brothers[26] put forward the concept of 'triage', borrowed from the military strategy of allocating medical aid during a war, as the basis for Western aid policy. On this argument the developing nations should be divided into three categories depending on their chances of 'survival'. 'Aid' should be restricted to countries like Pakistan and Tunisia which could possibly be 'saved' given food and development aid, but this should be made conditional on the implementation of appropriate population policies. The 'walking wounded' (countries like Gambia and Libya) could probably become self-sufficient under their own resources, whilst the final tragic category – the 'doomed to die', including Haiti, Egypt and India, have no hope of survival no matter how much aid is given and should be left to their fate.

Overseas 'aid' has of course always been used as a political lever to manipulate the economies of the developing countries. What is new is that this should be so openly advocated, with population control the central focus. Garret Hardin takes the argument even further:

> In a less than perfect world, the allocation of rights based on territory must be defended if a ruinous breeding race is to be avoided . . . It is unlikely that civilisation and dignity can survive everywhere; but

> better in a few places than in none. Fortunate minorities must act as the trustees of a civilisation that is threatened by uninformed good intentions.[27]

With the accelerating economic crisis in the West at least one group of such self-appointed 'trustees' have taken the opportunity to mount a new attack on 'uninformed' liberal values and advocate a new 'closed doors' policy in international aid. One series of full-page advertisements entitled 'The real crisis behind the food crisis', placed in many major newspapers and magazines a few years ago, was signed by such luminaries as Zbigniew Brzezinsky (President Carter's National Security Adviser), Isaac Asimov, Paul Ehrlich, Garret Hardin, Paul Getty, C. W. Cook (Chairman of General Foods Corporation), Edward Dwyer (Chairman of ESB Incorporated), William Phillips (Chairman of Multifoods Corporation), Burt Goodkin (Vice-Chairman, H. J. Heinz & Co.) and Henry Luce (Vice-President of Time Inc.) amongst many other representatives of US academia and big business. It declared:

> The world as we know it will likely be ruined before the year 2000 . . . world food production cannot keep pace with the galloping growth of population . . . 'Family Planning' cannot and will not, in the foreseeable future, check this runaway growth . . . the momentum toward tragedy is at this moment so great that there is probably no way of halting it . . . It makes no difference whatever how much food the world produces, if it produces people faster . . . Exponential population growth is basic to most of our social problems . . . of inflation, unemployment, food and energy shortages, resource scarcities, pollution and social disorder.

The statement concludes:

> There can be no moral obligation to do the impossible. No one really likes triage – the selection of those nations most likely to survive and the concentration of our available food aid on them . . . at some point we in the United States are going to

> find that we cannot provide for the world any more then we can police it . . . We must not permit our aid to underwrite the failure of some nations to take care of their own. When aid-dependent nations understand that there are limits to our food resources, there is hope that they will tackle their population problems in earnest.[28]

The ideology of population

Why should it be that populationist ideas are so dominant within the environmental movement in the capitalist nations? One reason is the attractiveness of an argument which *reduces* a complex of interacting problems – poverty, malnutrition, resource depletion and environmental pollution – to a single 'cause' – too many people – and which is apparently supported by the 'facts'. The world population is currently growing at around 2 per cent per annum, which implies a doubling over the next thirty or so years, whilst world hunger and ecological crisis seem to become daily more acute.

In fact, of course, such 'facts' are highly selective. Since 1945 (and indeed, as far as can be ascertained from the available data, since the time of Malthus) world food production has grown faster than population. There has never been a year nor a geographical region when per capita production of either protein or calories have fallen below FAO mean minimum levels. The problem, then, is one of the unequal *access* to resources, and the inability of the poor to purchase them, rather than simply one of 'too many people' (or of insufficient production). In fact, studies show that existing levels of technology and best agricultural practice, properly applied could produce vastly more without damage to the environment – sufficient to feed several times the present world population (although this is of course no reason to argue *for* continued population growth). There is presently a gross underproduction as well as maldistribution of food. In the USA some 20 per cent of cropland (around 75 million acres) lies fallow under Reagan's policy of paying farmers not to produce in order to keep prices high. Common Market stockpiles of grain are some five times larger than the amounts required to provide the FAO with a supply to meet the immediate needs of humanity. In addition to unfarmed land and stockpiles maintained to protect profits, and in addition to the gross waste of primary crops for brewing and for animal feeds (much of it imported as high-grade protein such as soybean or fish-meal animal feed supplements from the Third World) in the industrialised countries, there is a tragic misuse of land in the ex-colonial countries themselves. In Columbia for example, over 70 per cent of the fertile land is owned by large landowners, most of it untilled, and in the whole of Central America and the Caribbean 70 per cent of children are estimated to be malnourished whilst 50 per cent of the agricultural land is producing export crops like tomatoes or cut flowers for North America.[29]

At the same time, many countries and regions can point to quite large drops in the fertility rate where family planning campaigns have been coupled with progressive socialist policies. For example, China has over one billion of the world's 4.5 billion people, yet nobody talks any more of China's 'starving millions'. Moreover the population growth rate has been reduced since the early 1970s from 2.3 per cent to 1 per cent. As the first edition of the Brandt report pointed out:

> Even very poor areas such as Kerala in India have been able to give people new hope for a better life by involving them in the workings of development, improving their health, raising the status and educational levels of women as well as men, and ensuring adequate food supplies for the poor . . . birth rates have fallen while they remain high in richer developing countries which have paid less attention to the needs of the many.[30]

What the Brandt Commission omitted to mention is that Kerala has a communist government. But even in distinctly non-socialist countries like Chile, Columbia and Costa Rica (the first countries in Latin America to systematically introduce birth control facilities) birth rates have fallen by over a third in the last two decades, as they have (for different reasons) in Hong Kong, Singapore and South Korea. The most recent World Fertility Survey found that decreasing birth rates characterised almost all Third World countries except Africa, and some of the poorest Asian countries (like Bangladesh and Pakistan).

As the 1974 World Population Conference declared, 'development is the best contraceptive'. What the developing countries need is economic growth, better agricultural systems, educational health and social services and a new status for women. This implies development along *socialist* lines, with technologies consonant with their ecology and particular social and economic needs.

The involvement by capitalist governments in the economic policies of most Third World countries is to encourage quite a different sort of development however – 'development' to provide resources, markets and profits for multinational companies. And far from being a philanthropic gesture from the 'rich' to the 'poor' countries of the world, much international 'aid' is primarily designed to increase the political domination and control of multinational capital, and to frustrate those very developments which (amongst other things) would permit a reduction in rapid rates of population growth. Yet ironically, rapid population growth in Third World countries is increasingly perceived as a threat to their 'stability' and continued exploitation. This 'threat' has been a major influence on the attitudes of Western governments, in particular the USA which, with just over 5 per cent of the world's population, consumes well over 50 per cent of the world's resources. This is how Lyndon Johnson clearly saw the purpose of US 'aid' to Latin America as a substitute for social and economic reform. 'Let us act on the fact that less than five dollars invested in population control is worth a hundred dollars invested in economic growth.[31]

In a revealing comment, Robert McNamara while President of the World Bank (from 1968 to 1981) stated that its activity on population matters:

> arises out of the concern of the bank for the way in which the rapid growth of population has become a

> major obstacle to social and economic development in many of our member states. Family planning programmes are less costly than conventional development projects and the pattern of expenditures involved is normally very different. At the same time we are conscious of the fact that successful programmes of this kind will yield very high economic returns.[32]

It was Robert McNamara who as US Defense Secretary in the 1960s, prior to his term at the World Bank, attempted to destroy one struggling Third World country when he ordered the bombing of Vietnam. He was also one of the instigators of the Brandt report.

Population: a non-Malthusian perspective

The popular acceptance of 'populationist' arguments arises not just from their superficial plausibility but also from the fact that they put the blame for world poverty and economic and environmental problems squarely at the door of the poor – specifically on their reproductive organs. At the same time, 'it is clear that the fact that we name the interests which lie behind current demographic theories will not conjure the needs of a rapidly growing population out of existence.'[33]

It has always been the arrogance of the affluent who imagine that they have the solutions to the problems of the poor. Most people are not fools, and generally make rational decisions relating to the immediate situation they find themselves in. If poor peoples in the Third World have large families at least one reason for this is because it is advantageous for them to do so. As one Indian blacksmith expressed it:

> There is only one way out. And that is to have enough sons. Don't smile. If I have sons they will work outside, labour even, as animals do; but save . . . A rich man invests in his machines. We must invest in our children. It's that simple.[34]

In a society where poor sanitation and medical facilities mean a child's chance of survival to maturity is uncertain, and where welfare provision is non-existent, a large family often means the only chance of a moderate level of security in old age. A 1968 survey in Pakistan showed (using language that reveals some crucial assumptions about the roles of women) that a woman needed to have 'at least 6.3 children to achieve a 95 per cent probability of having at least *one son when her husband reached 65*' [emphasis added].

In these circumstances, where for a multitude of reasons a small family is a grave economic and social disadvantage, it is little wonder that birth control campaigns, including 'family planning' education and exhortation, incentives in the form of transistor radios in exchange for vasectomies, or coercion like that of the Indian sterilisation programmes, have met with little success.

If anything, rapid population growth should be seen as a *consequence* rather than a *cause* of poverty, malnutrition, and a degraded environment. As one commentator has put it:

> The solutions put forward by the population theoreticians generally lay the major emphasis on birth control. All such schemes will fail. Birth control merely treats a *symptom* and leaves the root causes untouched . . . Talk of overpopulation directs attention away from the root causes of these problems. Birth control itself will solve none of them.[35]

Barry Commoner takes this even further:

> So long as the total resources appear to be sufficient the approach (of trying to reduce population) . . . is equivalent to attempting to save a leaking ship by lightening the load and forcing passengers overboard. One is constrained to ask if there is not something radically wrong with the ship.[36]

And that, of course, is a *political* question.

One route to an answer is to examine the different demographic patterns within the industrialised nations and within the Third World in terms of the historical and continuing economic and political relationships between the two. If one takes the demographic history of an advanced industrial country such as the UK then a distinct pattern emerges.

The different demographic phases are each associated with a stage in the historical, social and economic development of the UK. The first stage, with both birth and death rates rising, is associated with the early phase of industrialisation. The second, from the mid-eighteenth century to the end of the nineteenth, is characterised by a constantly declining death rate whilst the birth rate remains static at a high level. This period is one of steady expansion of capitalism, the growth of industry and productive capacity, the rapid expansion of towns and cities, and most importantly, the acquisition of colonies, overseas exploitation, and the growth of empire. In the third stage, the processes (economic, social and demographic) begun in the first and developed in the second stage begin to come to (for want of a better word) maturity. The death rate levels off, but the birth rate falls steeply. These processes are to some extent self-accelerating since every fall in the birth rate changes the age structure of the population – increasing the proportion of people in the high age groups, and increasing the death rate. Since proportionately fewer women are in their major reproductive period (under thirty years) the fall in the birth rate occurs even faster.

Changes in population are primarily a function of the difference between birth and death rates and the four phases described are thus reflected in the pattern of population growth. During the first period of high mortality and high fertility the population grows steadily – but slowly. During the second, rapidly falling death rates whilst the birth rate remains high lead to an exponential increase in population – an 'explosion' analogous to that which can be seen today in many developing countries. In the third, with the birth rate falling, the rate of population growth itself falls and the population whilst still growing, does so much more slowly than before until, finally, it begins to 'level off' and to reach a state of equilibrium between births and deaths. The whole process has come to be known as the *demographic transition* and seems to be characteristic of all developed countries. In Europe, for example, the precise pattern varied from country to country but most rates of population growth are well under 0.5 per cent implying a 'doubling time' of more than a century and a half.

Turning to patterns of population growth in the developing countries, it can be seen that entirely different forces are at work, and that the problems of population stabilisation are likewise entirely different. Typically, these countries have rates of population growth of well over 2 per cent per annum, implying 'doubling times' of under thirty years, with all that this involves in terms of pressure on already overtaxed (and still grossly inadequate) resources.

Rich and poor: two worlds or one?

Since the 1950s there has been a considerable vogue for explanations of changing population structures in terms of the 'demographic transition' a concept originally put forward in 1945. 'Transition theory' is essentially an empirical generalisation based upon the observed population history and levelling off of rates of population growth due to falling birth rates common to all industrialised societies. However, it has attained the status of almost a general 'law'. The implication is that the developing countries are 'merely' in the first stages of their own 'demographic transition' – the stage that the UK was in, for example, a century and a half ago – and that in due course the 'transition' will be completed. In this latter sense transition theory has been eagerly seized upon by many socialists and counterposed to the traditional Malthusian analysis, the argument being that what developing countries need is not 'checks' to their populations (whether in the form of family planning campaigns or the more positive 'checks' of starvation or coerced sterilisation) but rather 'development' which will allow them to undergo their own demographic transition as the developed countries have done before them. However this ignores the *causes* of underdevelopment and of the 'population explosion' both of which are the consequences of the historic and continuing relationship between 'rich' and 'poor' nations.

The origin of underdevelopment

Demographic processes in the developing countries are not simply similar-but-later parallels to what has already happened in the West, any more than underdevelopment itself is simply a 'failure' on the part of some countries to undergo an industrial revolution that started in the UK two centuries ago. As André Gunder Frank has shown, underdevelopment is itself a *consequence* of the development of Western capitalism, and its subsequent world hegemony.

> Even a modest acquaintance with history shows that underdevelopment is not original or traditional and that neither the past nor the present of the underdeveloped countries resemble in any important respect the past of the now-developed countries. The now-developed countries were never *under*developed though they may have been *un*developed.
>
> It is also widely believed that the contemporary underdevelopment of a country can be understood as the product or reflection solely of its own economic, political, social and cultural characteristics or structure. Yet historical research demonstrates that contemporary underdevelopment is in large part the historical product of past and continuing economic and other relations between the satellite underdeveloped and the now-developed metropolitan countries.[37]

Colonialism *created* underdevelopment,[38] just as in many countries today neo-colonialism, the economic domination by foreign multinationals and political pressure by their Western governments, are major obstacles preventing development along a democratic socialist path appropriate to the ecology of each country and the needs of its peoples. Precisely the same argument can be applied to demographic change. The development of modern capitalism has itself depended largely on the extraction of raw materials and the exploitation of labour in its colonies and economic 'dependencies'. As the Brandt report reflects, some sections of capital itself see future prospects for profits lying increasingly in the development of Third World markets. Certainly the increased living standards towards the end of the last century and at the beginning of this, from which the last stage of the UK's own 'demographic transition' can be dated, coincided with the heyday not just of 'industry' but also of 'empire'.

The relationship has also worked the other way around; population growth in many areas of the underdeveloped world is associated directly with the history of imperialism and the export of new technologies. The current explosion of human numbers is a direct consequence of the continued domination of the developing world by the West.

Demographic parasitism

Several studies have documented the impact of imperialism on the demography and ecology of the exploited country. For example, the population explosion in Indonesia was set off by the introduction of new technologies and living conditions by the Dutch. There is evidence that the birth rate in the colony was consciously fostered by the Dutch in order to provide a growing labour force needed to exploit the colony's natural resources. The wealth extracted from Indonesia ended up in the Netherlands where it supported the Dutch through their own demographic transition. As Clifford Geertz points out:

> In effect, the first, or population-stimulating phase of the demographic transition in Indonesia became coupled to the second, or population-limiting phase of the demographic transition of the Netherlands – a kind of demographic parasitism.[39]

Subsequently, of course, continued economic exploitation has held back and distorted Indonesia's own economic and social development. Direct military involvement by the

USA as well as all sorts of indirect political pressure on the part of foreign capital, have been deliberately directed to frustrate attempts to engineer fundamental structural changes in Indonesia that could have provided for better living standards for the people and, *inter alia*, removed some of the conditions and pressure that make for rapid population growth. Moreover, with the post-war development of synthetic chemicals, Indonesia's natural rubber trade declined, thus further depleting the economic opportunities for advancement that might support their own motivation for population control.[40]

A great deal of work remains to be done in the study of the relationship between demographic processes in the capitalist countries and those in their colonies, but it is likely that in many cases the same sorts of 'parasitic' processes can be shown to exist. That is, the 'demographic transition' of the industrialised nations is not some sort of 'automatic' process occurring in these countries in isolation. It is rather a phenomenon specifically related to the quite different demographic patterns of Third World countries in the same way as 'development' and 'underdevelopment' are two mutually consequent sides of the same imperialist coin.

The 'blame' for rapid world population growth cannot be laid at the door of the peoples in Third World countries, nor attributed to one unfortunate but accidental predicament that poor countries happen to find themselves in. The 'blame' – and hence a major proportion of the responsibility for a solution – lies fairly and squarely on the shoulders of the developed capitalist nations themselves.

Population and progress

It is precisely the economic stranglehold of multinational capital and the anarchy of the international marketplace, together with the international and 'development' policies of Western governments – who in the last analysis always act in their own interest – that are the major obstacles to social and economic change in the Third World. And, it should by now be clear that only such change can begin to solve the pressing environmental and social problems in these countries – including the problem of rapid population growth. Despite the professed concern of the Brandt Commission, of the Club of Rome, of the Rockefellers and McNamaras for the peoples of the Third World and the future of our planet, despite all the liberal humanist rhetoric, the real issue for capital is the wealth which it can continue to extract from this planet and its peoples.

Much of the concern expressed in the West over the 'population explosion' in the Third World is at best misguided, at worst motivated by self-interest or the desire to conceal the real causes of the world 'environmental crisis'. Rapid population growth is, of course, a problem – for the newly socialist developing countries it is one of the most intractable problems they face in the short term and a constant obstacle to efforts to improve the conditions of life for the mass of the people. But on a world scale it is not *the* problem, and a historical and political analysis shows rapid population growth to be – ultimately – a consequence rather than a cause of poverty. In a nutshell, it is the historical product of international capitalism which must be destroyed if any long-term solutions are to be found. If individual countries – and indeed our planet – are being taken closer to the brink of ecological disaster, it is not by 'too many people' nor by any such 'singular fault, which some clever scheme can correct, but by the phalanx of powerful economic, political and social forces that constitute the march of history. Anyone who proposes to cure the environmental crisis undertakes thereby to change the course of history.'[41]

Ecology can be an important weapon in this struggle, but like all weapons it can be used to help people, or hinder them. 'Ecological warfare' has already been used against many Third World peoples and in the hands of big business today ecology is being used to make profits rather than to improve the lives of people. It is also necessary to challenge the ideology of ecology and to transcend the analysis of ecologists like Colvinaux who, at the end of one widely-read US textbook, offers the student who asks 'what can I do to help' the advice: 'Make sure you have no more than two children yourself and try to persuade your friends to do the same.'[42]

'Changing history' should be the job of all those who care

about the future of our species on this planet. It is a process that has already been started – by the peoples of Angola, of China, of Cuba, of Vietnam and Zimbabwe, and of all Third World peoples who have thrown off the burden of imperialism and begun to build a new society. It is a struggle already being waged by peoples in Brazil and Chile, in Sri Lanka, India and Pakistan, in South Africa, and indeed in every country of the world. Our job is to join it.

9. Ecology, interspecific competition and the struggle for existence

Jonathan Silvertown

Introduction

If you meet that rare animal, the person in the street, and say you are a biochemist, a geneticist or even a biologist, most people will look at you blankly, but if you say you are an ecologist you may find that a light dawns in their eyes. Many people have a vague idea about ecology and it usually conjures up things such as pollution, whales, the balance of nature, overpopulation, Friends of the Earth, and so on. There are many ecological political parties in Europe and ecological movements exist in most parts of the First World. In spite of the fact that most of the words that come out of a mental free association with the word 'ecology' are not found in ecology textbooks, there are, nevertheless, firm connections between the academic discipline and the popular conception of ecology.

These connections are the first clue to the social influences which have operated on the subject during its development. The first and arguably the strongest connecting link is the concept of scarcity. It is a part of common-sense 'ecological wisdom' that the supply of fossil fuels is limited and that its use should be regulated; it is supposed that food and living space are also limited and hence it has often been argued that the human population should be controlled (see chapter 8). Scarcity is also a key concept in academic ecology. The largest sustainable population of an animal or a plant in an area is called the carrying capacity of the environment. If exceeded, a scarcity of resources such as food will cause mortality which will return a population to the carrying capacity or cause a population crash.

Another link between popular and academic ecology is the idea of connectedness. Changes made to one part of the environment will automatically have consequences for other parts which may also change. Hence, ecological catastrophes may follow from seemingly insignificant acts of pollution or interference. There are various opinions on connectedness amongst academic ecologists but most would acknowledge that complex interactions between different organisms do occur. Some ecologists have coined the term 'keystone species' to refer to a crucial member of a community of organisms. Remove or perturb the keystone species and all the rest come tumbling down.[1]

The third link is homeostasis or 'the balance of nature': the environment appears to be able to withstand some perturbations but it can be pushed too far. Small clearings in tropical forest may be recolonised by trees and will return to forest. Very large clearings such as those being made along the trans-Amazon highway in Brazil may never recover. Academic ecologists disagree about just how resilient natural ecosystems are[2] and to what extent they compensate for perturbations by 'corrective' changes. But the fact that this debate is taking place at all shows that there is common ground between academic and popular ecological ideas.

The three links mentioned show that there is a trade in ideas between academic and popular ecology. This trade is two-way, and it is longstanding. To go to the roots of present-day ecological ideas, we need to trace back to the late eighteenth century, concentrating on the important concept of scarcity and the associated idea of competition.

The Great Chain of Being: nature in the eighteenth century

Eighteenth-century zoologists and botanists were the heirs to a longstanding conception of how nature was organised. They worked within a paradigm (or a world view) which placed all living, non-living and celestial creation in an hierarchical order, or the Great Chain of Being[3]. At the lower end of this were rocks and minerals, microscopic organisms and plants; next came simple invertebrate animals; and then vertebrates which were also ranked into a hierarchy from the simple to the more complex. Human society contained its own hierarchy and over it hung a heavenly hierarchy topped off by the Maker himself. The Great Chain of Being was believed to be continuous and consequently missing links were sought to fill gaps in the chain. When the small sessile animal *Hydra* was observed in 1739 it was hailed as the missing link between animals and plants.

Although the original conception of the Great Chain of Being was of a static hierarchy, with all of its stations occupied and every link fully forged, both of these ideas began to be challenged in the eighteenth century. The static chain denied that either extinction or the appearance of new species was possible and this was challenged by many people. Dr Johnson for example attacked the Great Chain of Being on logical grounds, but the most serious challenges to the status quo came from France. Voltaire described his own disenchantment with this notion:

> At first the imagination takes a pleasure in seeing the imperceptible transition from inanimate to organic matter, from plants to zoophytes, from these to animals, from these to genii, from these genii endowed with a small aerial body to immaterial substances; and finally angels, and different orders of such substances, ascending in beauties and perfections up to God himself. This hierarchy pleases those good folk who fancy they see in it the Pope and his cardinals followed by archbishops and bishops; after whom come the curates, the vicars, the simple priests, the deacons, the subdeacons; then the monks appear, and the line is ended by the Capuchins.[4]

This scheme, so convenient to the church, did not exist in nature.

At the end of the eighteenth century the French Revolution was an event which haunted the British ruling class as a terrible warning of what might happen if the established order was allowed to slip. Whilst it inspired radical ideas in every intellectual sphere in France, it also brought forth a reactionary defence of the social order in Britain. Although he was himself later a victim of the French Revolution, le Marquis de Condorcet wrote an essay claiming the possibility that under suitable conditions, where misery and suffering were abolished, human life could be indefinitely extended. He argued that all distinctions based upon race, class and sex could be swept away and that humanity was capable of improvement to perfection. A little later, in the sphere of biological ideas, Jean Baptiste Lamarck advanced a parallel argument that animals were also capable of change through gradual improvements in their structure and habits. In the revolutionary conditions of France in the late eighteenth century all beings, including animals, were to be set free from their allotted stations in the hierarchy ordained by God and so carefully preserved by his self-appointed temporal rulers.

Scarcity and the defence of the social order

An English cleric, the Revd Thomas Malthus, published in 1798 a book entitled *An Essay on the Principle of Population as it Affects the Future Improvement of Society, with Remarks on the Speculations of Mr Goodwin, M. Condorcet, and Other Writers.*[5] He argued that human society was incapable of improvement and he supported this with an argument which has had many intellectual and practical repercussions since (see chapter 8). His theory explained misery and vice as consequences of a natural law (the pressure of population on resources) and provided theoretical reasons why society could not fundamentally change. Although Malthus described this predicament and explained it in scientific terms, it appears he also believed that Christians should try to alleviate suffering, though they could not to try to reverse natural law (which was impossible anyway) but, in modern parlance, to give it a human face. Like many modern academics who provide biological underpinning for the status quo, Malthus distinguished between the act of describing the way things are and the act of subscribing to the 'regrettable' results of a natural law he had discovered. This sophistry hides the fact that a scientist begins an enquiry with preconceptions and a view of the world which must, to some degree, be reflected in any theory he or she later produces. Would Malthus have been so interested in refuting Condorcet in the first place if he had been a shoemaker instead of a clergyman? Why did he interpret his law as a conservative process tending to preserve things as they are rather than as a motive force for change? We will see shortly that the latter was indeed the interpretation that later scientists put on Malthus' law.

Just as Malthus argued that major improvement was impossible in human society, so the rest of the creation in the late eighteenth century and early nineteenth century was viewed as static and unchangeable by the clergy and many biologists. The interdependence of biology and theology, and both of these and the social order, was reinforced in early nineteenth-century England by a peculiarly English blend of science and theology, consolidated by Revd William Paley in 1802 with his book *Natural Theology*.[6] In it he put forward the argument that the intricate mechanisms found in animals and plants and the manifest adaptation of form to function in organs such as the eye, was evidence of design. A design supposes a designer, that designer must be a person and that person is God. Thus all evidence of adaptation was evidence of the existence of God and the close association of an adaptation with its function was evidence of the wisdom and goodness of God. This line of reasoning, known as the argument from design, spawned a whole school of biological treatises which documented morphological adaptations of form and function as a witness to the hand of the Creator. Because the argument from design depended upon a close fit of form and function, changes in morphology – in other words evolution – were a heresy which challenged the existence and goodness of God. If perfection already existed in the Creation, and had existed since the Beginning, there was no room for change or evolution.

Various theories of evolution did, however, challenge the fixity of species and hence all that the British establishment associated with it and had built upon it. Before Darwin published the origin of species, the most popular with the reading public was *Vestiges of the Natural History of Creation*, published anonymously, by Robert Chambers in 1844.[7] It met with a storm of protest from the establishment which

Darwin observed with trepidation, having already formed but not published his ideas on evolution.

One of the key ideas in the *Vestiges* and in other pre-Darwin evolutionary theories was that evolution was progressive – in other words that changes and developments in the living world were driven by some metaphysical force towards greater and greater organisation and perfection. Chambers and others interpreted the Great Chain of Being which was traditionally conceived as a static hierarchy as a chain of evolutionary development. Each link in the chain was an evolutionary stage on the road to perfection.

Scarcity as the motor for evolutionary change

In 1852 Herbert Spencer, an English philosopher of great importance in his time, united the ideas of Malthus on population and progressionist evolution into a theory of social evolution. This was set out in two articles 'A theory of population, deduced from the general law of animal fertility' and 'The development hypothesis'.[8] His argument was that the pressure generated upon the human population by a scarcity of resources had a beneficial effect upon human society because it encouraged competition and selected the best from each generation. Spencer attributed the progress of human evolution from the earliest times to this process which he named the 'survival of the fittest'. Today this phrase is usually associated with Darwin but he did not use it in the first edition of *The Origin of Species* and only later introduced it on the suggestion of Alfred Wallace.

Spencer developed a 'synthetic philosophy' in which he attempted to subsume all branches of human knowledge.[9] A major part of this system of philosophy was a general law of evolution which attempted to show evolutionary processes in sociology, ethics and in the inorganic world as well as the organic. Spencer's faith in a natural law of progressive evolution led him to the same conclusion as Malthus, that scarcity and the misery it produced in human society were parts of a natural process which could not be altered. In fact Spencer went further than Malthus and maintained that attempts to interfere with evolution were dangerous. He opposed state aid to the poor on the grounds that 'the whole effort of nature is to get rid of such, to clear the world of them and make room for better'. The irony of Spencer's use of Malthus was that he maintained that evolution, driven by the effects of scarcity on population and by the survival of the fittest would *inevitably* produce human perfection. Malthus' original reason for writing his essay on population was to prove the opposite.

Spencer's belief in the hidden hand which guided universal evolution was so complete that he advocated the abolition not only of poor laws but state education, public health measures, taxes, state banking and a government postal system. He subscribed to the philosophy of laissez-faire to a degree recently matched by Milton Friedman, Margaret Thatcher and Ronald Reagan. All are inspired by the eighteenth-century Scottish economist Adam Smith whose book, *The Wealth of Nations* published in 1776, is the bible of laissez-faire economics.[10]

In the nineteenth century the departments of knowledge (as they would have been called) were not nearly as rigidly separated as they are today. An amateur, Robert Chambers felt confident enough to write a book on evolution which drew upon astronomy, zoology, botany, geology and palaeontology. Herbert Spencer explicitly set out to unify all human knowledge with his synthetic philosophy. Darwin wrote on a large number of subjects which few biologists today could match for scope. Particularly in the second half of the nineteenth century, there was also a large public for popular scientific works. In this kind of atmosphere social and biological ideas are not easily separated.[11]

Darwin's studies during his voyage on the *Beagle* had convinced him that species were not fixed entities, and in July 1837, six months after his return he began to keep the first of his notebooks on the 'transmutation of species'. In his autobiography, Darwin wrote:

> I soon perceived that selection was the keystone of man's success in making useful races of animals and plants. But how selection could be applied to

> organisms living in a state of nature remained for some time a mystery to me.[12]

Then Darwin records that in October 1838:

> I happened to read for amusement Malthus on population, and being well prepared to appreciate the struggle for existence which everywhere goes on from long-continued observation of the habits of animals and plants, it at once struck me that under these circumstanes favourable variations would tend to be preserved, and unfavourable ones to be destroyed. The result of this would be the formation of new species.

Darwin was not to publish his theory of the origin of species by natural selection until twenty years later when Alfred Russell Wallace wrote to him asking his opinion of an essay entitled, 'On the tendency of varieties to depart indefinitely from the original type'.[13] It contained a theory of evolution identical to Darwin's. Wallace had arrived independently at the same theory as Darwin, and Malthus' essay of population had played just the same part in his own invention of the idea.

Social Darwinism

Spencer, Darwin and Wallace all derived the idea of a struggle for existence from reading Malthus. The intentional effect of Spencer's theory was to legitimise competition and the struggle for existence in human society. Neither Darwin nor Wallace subscribed to Spencer's laissez-faire indifference to suffering or to his faith in the hidden hand of progress. Nevertheless the theory of evolution by natural selection was a product of the values of nineteenth-century society with its emphasis on competition; the subsequent interpretation of the theory was also connected with those values.

The theory of evolution by natural selection was vigorously opposed by the clergy when *The Origin of Species* was first published in 1859. A chief objection was that it was a materialistic philosophy which left no role for God in the world. Darwin himself did not defend his theory in public and this task was undertaken by Thomas Huxley. Despite the furore from the clerics, the theory gained rapid acceptance amongst scientists and intellectuals. Darwin's theory then began to be adopted and developed by social theorists, and thus social Darwinism was born.

Herbert Spencer's evolutionary scheme now had the solid foundations of Darwin's work to build upon. The view that individual success in human society was the result of the survival of the fittest in a struggle for existence became a popular notion, particularly with those who had acquired wealth and position.

> The growth of a large business is merely a survival of the fittest . . . This is not an evil tendency in business. It is merely the working-out of a law of nature and a law of God.
> (John D. Rockefeller (c.1900))[14]

> A struggle is inevitable and it is a question of the survival of the fittest.
> (Andrew Carnegie (1900))[15]

> The fortunes of railroad companies are determined by the law of the survival of the fittest.
> (James J. Hill (1910))[16]

These views were not confined to millionaires but prevailed in more subtle form amongst others too. Thomas Huxley wrote in *The Struggle for Existence in Human Society* that the Hobbesian war of each against all was the normal state of existence during human evolution.[17] However, Huxley did not subscribe to laissez-faire and considered that economic and social affairs should be regulated in order to make Britain competitive in the economic struggle for existence with other countries. To this end Huxley advocated that workers wages must be restricted in order to keep manufacturing costs down.

Many late nineteenth and early twentieth century writers

carried social Darwinism to its logical and brutal conclusion, regarding other races and even other European national groups as doomed in the struggle with the superior Englishman. John Louis O'Sullivan wrote in 1885 in an article in *Harper's* Magazine[18] that, 'the day is at hand when four-fifths of the human race will trace its pedigree to English forefathers'. In a picture album of the kind many US families must have possessed at the turn of the century, and which indicates the view of the world presented to untravelled Americans by journalists at the time, is a photograph of 'Natives of Queensland, Australia'. Beneath this photograph is the caption:

> The sad fact impresses itself upon the traveller as he looks upon these millions of aborigines in the South Pacific, that little can be done to improve them. They are doomed. If left alone, and to the gradual development of centuries, they might perhaps evolve themselves into a higher order of humanity. But like the North American Indians, they are disappearing rapidly before a new and sturdier race, and the islands and continents of the great southern ocean will soon be peopled by their conquerors.[19]

To recap on the nineteenth-century history of the idea of the struggle for existence, we have seen that it began as an attempt by Malthus to refute the possibility of social progress using a biological argument about population and the scarcity of resources. Herbert Spencer then used Malthus' same argument about population and scarcity to come to opposite conclusions and to support a theory of universal, progressive evolution. Darwin developed his biological theory of evolution by natural selection supported with copious data. This theory was developed in a social milieu where its premises of scarcity and struggle were already prevalent and the theory was rapidly applied to justify the excesses of expanding capitalism and colonialism. This was the intellectual and social background to the development of ecological theory in the twentieth century.

Ecology and the struggle for existence

The connection between ecology and the theory of evolution by natural selection lies in the struggle for existence. Darwin subsumed under this phrase all the interactions between organisms and between organisms and their environment which keep populations in check. Thus it includes *intra*specific competition and *inter*specific competition for scarce resources, the effects of the weather and physical environment and the effects of predators. Malthus emphasised the effects of *intra*specific competition. When Darwin realised the significance of Malthus' argument and began to formulate his own theory, he did not at first appreciate the full consequences of *inter*specific competition. In his autobiography[20] he says:

> At that time I overlooked a problem of great importance . . . The problem is the tendency in organic beings descended from the same stock to diverge in character as they become modified . . . The solution, as I believe, is that the modified offspring of all dominant and increasing forms tend to become adapted to many and highly diversified places in the economy of nature.

The evolutionary process of diversification Darwin refers to is now known as adaptive radiation. His 'places in the economy of nature' are the equivalent to the modern concept of the ecological niche. The finches observed by Darwin in the Galapagos are an example of a group of closely related species which are thought likely to have evolved from the same ancestral colonists, but which now, by adaptive radiation, have evolved different feeding habits.

Early twentieth-century community ecology

The general idea of a struggle for existence was applied constantly in social theory in the late nineteenth and early twentieth centuries. Interspecific competition was also studied by ecologists of the period but their work was little influenced by Darwinism and there was not the emphasis later placed upon it as *the* process of major importance in ecological communities. It is sometimes seen as strange that three pioneers in ecology, Tansley, Sukatschew and Clements, each carried out specific experiments on plant competition early[21] but then 'abandoned' this type of investigation and spent the rest of their careers studying whole plant communities and their dynamics. This may represent a shift in the research priorities they perceived but is more probably a modern misunderstanding of the theoretical framework within which they worked.

Studies of plant distribution and succession, both of which were of major interest then, incorporated competition as *one* mechanism amongst others that were relevant. Thus the large majority of the hundreds of ecological papers published between 1920 and 1959 which mention competition, do so in the context of species' distribution, succession or some problem in ecological management, and not as a subject in its own right. A few exceptions to this pattern, significantly, were published in the journal *American Naturalist* in which evolution was a dominant theme.[22] Ecologists of the period seem to have been practising a holistic science (or top–down, in the terminology of chapter 1).

I suggest that in the context of early community ecology, experiments involving competition were useful to exemplify its effects, but once the *potential* influence of competition had been demonstrated, it was no longer thought worth pursuing for its own sake. When the dominant ecological paradigm changed to a reductionist (bottom–up) one, competition was thrust to the surface as *the* most important ecological mechanism: one which had to be studied in its own right. Hence, to modern ecologists competition appears neglected before 1960, simply because today we are obsessed with it. The change from a top–down to a bottom–up approach seems to have been involved with the application of simplistic mathematical theory in ecology and the introduction of evolutionary ideas.

The switch to a reductionist paradigm

Three authors wrote upon the mathematical theory of competition, predation and the struggle for existence, but the names of Lotka, Volterra and Gause[23] did not achieve their present dominance in ecology until the 1960s when an entire edifice of theory was built, largely on the foundations of these three authors.

Gause wrote in his book, *The Struggle for Existence*,[24] that the Lotka–Volterra equation for competition between two species 'does not permit of any equilibrium between the competing species occupying the same "niche", and leads to the entire displacing of one of them by another'. In the same book Gause described some experiments that he performed on competition between species of *Paramecium*, in which the extinction of one species appeared to occur just as theory predicted. The methodology of these experiments has since been questioned; this particular result has, for example, been shown to be an artefact of the way in which Gause sampled the *Paramecium* populations in his experiment in order to count them.[25]

Many naturalists during the early twentieth century observed that closely related species occupying the same habitat appeared to avoid competition for food or space by occupying different niches. One of the examples Gause cited was a study by another Soviet ecologist, A.N. Formosoff, of a colony of terns composed of four different species: the sandwich tern, common tern, blackbeak tern and the little tern. The blackbeak tern was found to feed only on land, the sandwich tern fished in the open sea and the common and little terns fished near the shore but in water of different depths. Observations of this kind, reinforced by theory, led certain ecologists in the 1940s to adopt a principle which David Lack called Gause's principle. This stated that two species with similar ecology could not live for any length of time in the same region without one being excluded by competition from the other. When Gause's principle was discussed at a symposium of the British Ecological Society in 1944 it was recorded that 'a distinct cleavage of opinion revealed itself on the validity of Gause's concept'. An opponent of the concept, Captain Cyril Diver, listed many examples of related species which co-existed with no apparent ecological differences between them.[26]

Although Darwin did not state the competitive exclusion principle in any explicit way, he did draw attention to the point that the struggle for existence should be strongest between species in the same genus (congeners):

> As species of the same genus have usually, though by no means invariably, some similarity in habitats and constitution, and always in structure, the struggle will generally be more severe between species of the same genus, when they come into competition with each other, than between species of distinct genera.[27]

On the basis of this reasoning and employing Gause's principle, it is to be expected that species in the same genus will only rarely be found sharing the same habitat. Where they do share the same habitat, niche differences of the kind shown by Formosoff's terns should be detectable.

In a paper contributed to the British Ecological Society meeting in 1944, Elton[28] attempted to demonstrate the validity of Gause's principle by calculating the ratio of the number of species to the number of genera for plants and animals found in a number of habitats. His results showed that fewer congeneric species occurred together in small communities than in large ones. He deduced from this that interspecific competition between congeners in small communities was important in determining which species were to be found in them.

Elton did not take into account the difference in the ratio of species/genera which can be expected between large and small samples by virtue of the difference in the size of the two samples alone. A year after Elton's paper was published, Williams[29] tested Elton's data against a null hypothesis which allowed for the intrinsic effects of sample size on the species/genus ratio. He came to the opposite conclusion to Elton and found that congeners actually occurred together more often than was to be expected in small samples. This does not mean that species co-operate but it does suggest that the effects of interspecific competition are weak. It appears that similar species occur in the same habitat because they are adapted to similar conditions and that competition does not significantly reduce this tendency for congeners to aggregate.

Elton's test for competitive exclusion based upon the species/genus ratio of small communities was applied several times after Williams showed his method and conclusions to be faulty. Simberloff, who reanalysed these later studies, like Williams found that there were more pairs of co-existing congeneric species than was to be expected rather than fewer.[30]

At the time Williams's results had no influence against the view that the structure of ecological communities was shaped by interspecific competition. Gause's principle proceeded towards the status of dogma, becoming so firmly established that ecologists began to rewrite history and to ask why the importance of competition had not been appreciated earlier. The papers on succession and distribution which mentioned competition were forgotten or ignored.

A turning point came in 1960 for the fortunes of interspecific competition and the competitive exclusion principle. Hardin[31] suggested that although the idea was inherent in the early work of Darwin, Lotka and Volterra, there was no name for the concept so that it could not easily be discussed. This suggestion is not quite as ridiculous as it first appears. Hardin's own paper which appeared in the prestigious US journal *Science* conferred the impressive name 'The competitive exclusion principle' upon the concept. This act turned what was an *assumption* of Lotka, Volterra and Gause's mathematical treatment of interspecific competition into a dominant and inviolable rule. From a *postulate*, competitive exclusion was turned into *fact*.

Hardin further suggested that the competitive exclusion principle had been ignored for so long because people were squeamish about admitting that nature was red in tooth and claw. Cole[32] replied to Hardin's article in *Science* pointing out that alternative mathematical models to those of Lotka and Volterra (e.g. that of Skellam[33]) suggested that competing species could co-exist. He pointed out that the Lotka–Volterra model was unrealistic, that competition was not the only factor to be taken into account and that events in nature were far less deterministic than the model assumed. Cole argued that niche differences such as those observed in Darwin's finches occurred not because 'ecological differentiation is the necessary condition of co-existence' as Hardin and the exclusion principle claimed, but 'simply because natural selection will promote the spread of genes that permit a population to enlarge by exploiting an unfilled ecological niche'.

The next twenty years of research in ecology saw Hardin's view thoroughly explored and the skepticism of Cole and alternative theoretical models such as Skellam's virtually forgotten. It seemed as if competition was the only type of interaction between organisms that was worth attention. For example, mutualism – the virtual opposite of competition and an extremely common interaction – was completely neglected. In one review of mutualism, published in 1982, the authors felt justified in saying: 'The history of the study of mutualism by ecologists is akin to Sherlock Holmes' case of the barking dog, in which the point of interest was that the dog did not in fact bark.'[34]

A major stimulus to the development of competition-based ecological theory was given by an influential paper published in 1959 by G. E. Hutchinson, 'Homage to Santa Rosalia: or why are there so many kinds of animals?'[35] Although this paper was not exclusively about competition and niche separation,[36] one observation made in it led to a new attempt to formulate models of resource competition. A correlation often exists between the size of an animal, or the size of its feeding apparatus (jaws or beak), and the size of food it takes. Hutchinson suggested that a comparison of size measurements of this kind could be used to estimate the similarity in diet of two animals competing for the same type of food. Further, he proposed that competitive exclusion might only be avoided if the ratio of sizes was at least 1:1.4. This rule, later to become known as Hutchinson's ratio, was derived from some rather casual observations of co-existing species he had made. Theoretical ecologists rapidly developed a mathematical 'theory of species packing' which 'predicted' Hutchinson's empirical value for similarity, and a whole body of ecological theory was built upon the premise that competition was the primary determinant of species composition and of the diversity of communities.[37]

Field ecologists, impressed by the explanatory power and

rigour offered by the theory, began to search for other examples of Hutchinson's ratio. Often when an approximately 'correct' value was discovered this was interpreted as *evidence* of competition. Heck,[38] who first questioned the realism of the assumptions upon which the theory of species packing was based and showed that its predictions were sensitive to changes in these assumptions, had some difficulty getting his criticisms published. Several referees objected that existing theory was the only rigorous explanation for several observed patterns. What they did not question was whether those 'observed patterns' were really representative of patterns in the real world. We do not know how many field results were not published because they did not substantiate the theory of species packing and were therefore not thought interesting.[39]

A period of disillusionment with competition theory has recently set in.[40] It is now clear that ecological models of community structure based upon competition alone have their limitations.[41] Even where competition has been seen to explain some aspects of patterns of species distribution, other processes such as dispersal, environmental disturbance history and chance are now being acknowledged as contributory forces alongside competitive exclusion.[42] The assumptions of the old Lotka–Volterra equations are now being seen for what they are, just one way of looking at the world, and one which produces a far too simplistic view.[43]

De Bruyn, who recently published a simulation model which demonstrates the *co-existence* of competitors commented:

> Being, according to his own theory, a masculine descendant of apes, and having grown up in the capitalistic England of the nineteenth century, Darwin emphasised the role of competition in nature. Since then his ideas have become generally accepted by scientists, most of whom had the same sex, origin and social background.[44]

The impact of the prevailing economic ideology on how they see the world is often tellingly revealed when ecologists seek analogies to explain ecological theory. Diamond[45] addressing the same old question about why ecologists did not realise the importance of competitive exclusion earlier than they did, argues that they misconceived the nature of the struggle for existence.

> They looked around them to see individuals of different species but similar trophic roles fighting, rarely saw it, and concluded that interspecific competition was unimportant.

Diamond continues:

> Just imagine what errors you would commit if you were an economist and used the same reasoning . . . you would see Hertz and Avis counters adjacent at airports, would note that the ladies dressed in yellow were not fighting with the ladies dressed in red, and would conclude that Hertz and Avis do not compete. In fact, Hertz and Avis compete intensely for a shared resource, customers. But the mechanism of competition consists of trying harder for customers so as to starve out the rival's resource base, and not of fighting. Or you would note that RCA does not make computers today while IBM does, and you would conclude that competition is unimportant in the computer industry. Without having observed continuously through recent years, you would not realise that RCA once did make computers, but that IBM was so much more successful at harvesting customers without resort to fights that this competition ended in permanent exclusion of RCA from the computer industry viewed as a business habitat.

There is clearly some truth in de Bruyn's statement though perhaps it is a little unfair to Darwin. He *was* a man of his age, but in our century the 'struggle for existence' has been interpreted in a far narrower sense than his original conception justified. In social Darwinism it made the idea synonymous with commercial competition, eugenics and laissez-faire capitalism. In ecology, which began as a holistic science, the use of the reductionist method singled out interspecific competition which, for a time, seemed to be the only basis for theoretical generalisations in the subject. New theoretical models are now beginning to appear;[46] let us work for a new social order as well!

10. The tomato is red: agriculture and political action

Uriel Kitron, Brian Schultz, Katherine Yih, John Vandermeer

> The struggle for different technologies is essential to the struggle for a different society.[1]

Introduction: the wasteland

In modern agriculture, dominant technological patterns have permitted rapid expansion and greater productivity, but have not solved the problems of hunger and malnutrition. Furthermore, agricultural development is creating a wasteland of much of the world. The environment is being changed drastically through eutrophication, nitrogen oxides in the atmosphere, the loss of de-toxifying capacities in the soil, the spread of pesticides, and the destruction of wildlife. People's health is damaged by a lifetime of hard labour and by agricultural chemicals, which are poisonous, initially, to farmworkers and later to everybody. Food quality and nutritional value are subordinated to the requirements of market, storage, transport, and mechanical harvesting. Social inequalities have been intensified, as witnessed by widening differences in power between farmers and agribusiness, farmworkers and corporate farms, and underdeveloped agro-export countries and the developed world. Finally, agriculture's own productive base is being undermined through soil destruction and the loss of genetic diversity.

Some have maintained that the problem is technical, that the ecological and social waste laid by dominant agricultural practices results from inappropriate technology or from technology 'out of control'. Others argue that the problem is a social one and that progressive social change can prevent this destruction. But neither appropriate technology nor progressive social change, alone, can correct the course of agricultural development.

This chapter attempts to show a path out of the agricultural wasteland. We start by developing a brief historical framework. Beginning with the example of processing tomatoes, we show how technological change and political developments are intimately intertwined and how the dominant development pattern has produced a system where the productive social sectors are caught in a bind between suppliers and buyers. There is then presented a general theoretical analysis of the political role of technology, including tentative programmatic suggestions for radically altering the manner in which technology develops. Finally, with the stage set historically and theoretically, we describe two attempts at putting our theoretical position into practice, one in alliance with the revolutionary government of Nicaragua, the other in alliance with a union of militant farm workers in the USA.

Historical perspective: bitter fruit in the colonies

The tendency in agriculture has been for technology to use ever more machinery and energy per worker, a trend which derives both from the profit motive and from the associated need to control the labour force. This particular technological trajectory has led to dramatic increases in production capacity and unparalleled wealth for a certain minority. So impressive have been its successes that it is frequently regarded as the best of all possible trajectories, perhaps even as inevitable. While it has been remarkably effective in terms of profitability and control, its attractiveness diminishes when analysed more broadly, with different criteria of success.[2]

Because the early history of agricultural development is couched in a system that seeks profitability and labour control, certain constraints have operated to limit the possible trajectories along which it could proceed. In particular, those who own the land and the implements to work it have been in conflict with those who work on the land, except for those situations where the owner supplies all the labour (peasant producers or so-called family-labour farms). This conflict, operative since class societies have existed, is a clear influence in the history of processing tomatoes in the USA.

Prior to the second world war, tomato production in the USA was relatively static. Three general regions accounted for the bulk of production: an eastern zone (Maryland, New Jersey, New York, Pennsylvania, and Delaware), a mid-western zone (Michigan, Ohio, Indiana, and Illinois), and California. The three zones carved out more or less equal portions of the entire US market, with California slightly lagging behind – probably due to an inability to gain efficient access to the large eastern and mid-western markets. However, during the early war years several changes dramatically increased production efficiency in California.

The first and probably most important factor was the development of a new high-yielding variety (the Pearson tomato), which was particularly well suited to California growing conditions, allowing California's producers to increase production from an average of 6–7 tons per acre to as much as 20 tons per acre.[3] In addition, for a variety of reasons, farms in tomato-producing regions of California had become large enough to take advantage of economies of scale. Also, the federal government had targeted California for the production of tomatoes for the war effort, thus suddenly creating an almost insatiable market.[4] Finally, the use of pesticides became widespread.

These factors all combined to create ideal conditions for dramatic increases in production of processing tomatoes in California. But an important ingredient was lacking. The production of tomatoes had always been very labour-intensive. Industrial labour requirements associated with wartime production removed large numbers of men and women from the labour reserve that would have been used in the tomato fields. Thus, while the war effort helped to create a vast potential for expanded production in California, it also created the major stumbling block, a shortage of labour.

What was needed was cheap labourers, but furthermore, labourers who would not organise. Militant activity on the part of the trade unionists had been commonplace in California since the 1930s. In 1942 the first contract was signed with the Mexican government to provide workers for the California sugar-beet industry. This was the beginning of the 'bracero programme' in which thousands of Mexican workers were brought to the USA to provide a cheap, unorganised workforce for growers. The farm labour market in California was thus effectively protected against competition from the industrial sector, resulting in dramatically lower wages. While the bracero programme was initially an emergency measure to provide the necessary labour to meet the demands of the war effort, by the end of the war it became evident that fear of labour organising was a more important concern than the wartime shortages. Each year, under pressure from California growers, a contract was renegotiated with the Mexican government, until the programme was formally terminated in 1964.[5]

The full establishment of the bracero programme had two important effects, one in California and one in the Mid-west. First, the braceros provided the well-behaved labour necessary for California to take full advantage of new production technologies (e.g. the Pearson tomato) and begin its total domination of the tomato processing industry. Second, when the braceros flooded the labour pool in the Rio Grande valley of Texas, they reinforced the tendency of Chicano (US citizens of Mexican descent) farmworkers to migrate northwards to the mid-western states. The depressed farmworker wages in Texas, resulting from the bracero programme, made jobs in Ohio, Indiana, and Michigan look attractive. By 1960 more than ten thousand workers could be found in Ohio alone at peak harvesting season, the vast majority Chicanos from Texas. Thus, the mid-western tomato-producing region, while losing much of its share of the market between 1945 and 1965, gained a regular stream of migrant workers, mainly from Texas.

California's long history of migrant workers did not exist in the Mid-west. Consequently California's tradition of labour organising did not exist there either. Only relatively recently has a visible union been established in the Mid-west.

In 1968, farmworkers formed the Farm Labour Organising Committee (FLOC), with the immediate intention of organising tomato pickers in Ohio, but with the ultimate goal of organising all farmworkers in the Mid-west. Patterning their activities to some extent on organising in California, such as the United Farm Workers' successful struggle against grape and lettuce growers, they organised workers, called strikes, and won contracts. By 1974 they had negotiated over thirty contracts and were well established in north-western Ohio.[6]

However, tomato growers could not really afford to meet union demands because prices in the Mid-west are set by the processors that buy tomatoes from the growers, and growers cannot raise their expenditures on labour until they get a better price for their produce. Many growers simply began to shift production to corn and soybeans. Changing their strategy, FLOC began building a programme to bypass the immediate employer (the grower) and to put pressure instead on the true centre of political and economic power – the processing corporations. It was considered most practical to target only a few processors first; Campbell's and Libby's were chosen because they were the largest canneries in north-west Ohio where FLOC was organising. The programme comprised, first, a strike against all growers holding contracts with either Campbell's or Libby's canneries and, second, a consumer boycott of all Campbell's or Libby's products. The goal, of course, was to win a contract with the processors.

Among the various political acts employed by the processors in response to the strike and boycott has been the attempt to force the producers to mechanically harvest their tomatoes, thereby eliminating the need for labour altogether. Despite concerted efforts on the part of the processors and various state agencies, Ohio as a whole was only about 70 per cent mechanised in 1983 (about the same as 1979, the year after the strike was announced). The machine has not yet replaced the workers, and in this case possibly never will.[7]

The tomato industry is an excellent example of the interplay of political processes and technological development in agriculture.[8] Technological changes (e.g. the Pearson tomato) brought about a 'labour problem' (cheap well-behaved labour was not available to take advantage of the new technology). The labour problem was resolved through a political process (the bracero programme) which created other labour problems (increased Mid-west migrant stream, general lowering of agricultural wages, and eventually the unionisation of mid-western farmworkers). The new labour problems were thought to be resolved through a technology (mechanical tomato harvester), although a final resolution has not yet been reached in the Mid-west.[9]

As the interaction of these political processes and technological developments was shaping the tomato processing industry, it was also causing far-reaching changes in the social organisation of agriculture. The seed companies which market new varieties, the chemical companies which market the pesticides, and the companies that market the machinery all became part of the overall agricultural system. In combination with food processors, these input suppliers have grown to dominate agricultural production since the Second World War. At the turn of the century it made some sense to talk about the 'farmer' as the key figure in agriculture. Land and labour were combined to produce tomatoes. But since the Second World War, we have evolved into a system in which, metaphorically, agriculture is the production of ketchup from petrochemicals.[10] It is almost incidental that 'farming' is involved somewhere in that process.

As this process of heavy capitalisation proceeds, power relations also evolve. The general pattern has been that the input suppliers (machines, seed, chemicals) and the output buyers (retail groceries, food processors, grain elevators) have become inordinately powerful compared to the farmer. The farmer is thus caught in an economic squeeze between the suppliers who charge too much and the buyers who pay too little. This was the ultimate reason that FLOC was forced to seek contracts with the large buyers instead of the beleaguered mid-western tomato farmer.

A similar process has operated at the international level. Left relatively powerless as a legacy of colonialism, former colonies were frequently forced to pursue development on the basis of an agricultural economy. To modernise production and gain foreign exchange, they were forced to seek modern inputs and to sell their products to those very forces that had impoverished their economies in the first place. The agricultural sector as a whole may have been very profitable, but the profits were realised mainly by input suppliers and commodity buyers, both of which were located in the developed world. Profits were effectively extracted from the former colony and used to propel development in the more developed country, leaving the former colony in a perpetual state of underdevelopment.

Small agro-exporting countries, such as Nicaragua, thus find themselves in a position similar to the small tomato farmer. They are forced to pay exorbitant prices for their inputs and to sell their produce at relatively low prices, a pattern that was initiated with colonialism but dramatically reinforced with the modernisation trends following the Second World War. One of the major tasks facing underdeveloped agro-export countries like Nicaragua is to reduce their dependency on suppliers and buyers from the developed world.

As in other situations where power relations are contested, historical developments in agriculture have always pitted those who benefit from current arrangements against those who seek to change the world for the benefit of the majority. Food processors benefited when farmworkers remained unorganised, but the farmworkers obviously did not. FLOC and the conflict with Campbell's thus resulted. Somoza and his family benefited from the agro-export economy of Nicaragua but the people of Nicaragua obviously did not. The Sandinista popular revolution thus resulted. In all this, technological development is involved, not as the major determinant of political change, but in setting the stage for conflicts between conservative and progressive forces.

Transitional technology and a radical science programme

Technology is often assumed to be politically neutral, developing from previous technology as science (also assumed to be politically neutral) advances. Yet throughout the interwoven histories of society and technology, the dominant political interests and the design and use of technology have reinforced each other. Technology is not the product of some neutral, unidirectional progress. Current technology no doubt would look quite different if it had been developed in a different social milieu. Consider, for example, a society organised such that the costs of 'externalities' were calculated in the cost–benefit analyses of technology implementation instead of shunted off to workers and consumers – would the technologies that now prove to be cost-effective still pass the test? If the cost of pesticides were evaluated in terms of short- and long-term illness, eventual pest resistance, and the elimination of the pests' predators, would a heavy reliance on them still be regarded as an acceptable method of control?[11]

In reciprocal fashion, technology influences politics. Its political role comprises both material and ideological aspects:

> At a material level, technology sustains and promotes the interests of the dominant social group of the society within which it is developed. At the same time, it acts in a symbolic manner to support and propagate the legitimating ideology of this society – the interpretation that is placed on the world and on the individual's position in it.[12]

> Capitalism develops only those technologies which correspond to its logic and which are compatible with its continued domination. It eliminates those technologies which do not strengthen prevailing social relations, even where they are more rational with respect to stated objects. Capitalist relations of production and exchange are already inscribed in the technologies which capitalism bequeathes to us.[13]

Once we conceive of current technology as produced in the interests of the dominant social group, rather than the inevitable outcome of 'progress', we can begin to imagine technologies which would benefit other groups instead. Technology which helps change society in progressive ways we call *transitional technology*. What distinguishes transitional technology from alternative, appropriate, or traditional technology is the intent to transfer political power through its development and use. Essentially the same concept has been developed by Brazilian workers, who use the term 'appropriate technology'.[14] Alternative or appropriate technology, despite its obvious potential (to reduce dependency by employing local knowledge and natural resources, for example), can be co-opted by conservative interests.

Although it is clear that a technology deliberately designed to facilitate progressive social change is a transitional technology, no foolproof set of criteria exists to tell us what technologies are transitional. Technology's political impact depends on the political circumstances. It depends, further, on the conception held by the scientists involved of their role. If a scientist's aim is to develop transitional technology, she/he will be working from a political analysis; this analysis will reveal mistakes, cause the scientist to modify the course of the research or development, and make the development of truly transitional technology more probable. Whereas if the scientist does not think in terms of producing transitional technology, there will be no on-going political analysis and no refinement of the development to ensure that transitional technology results.

Recognising the political bias of technology and its potential for influencing social relations, progressive scientists are compelled to take sides. This implies certain actions:

(a) To actively ally with progressive forces. In order to develop transitional technology and otherwise make their work serve progressive interest, scientists must find out exactly what those interests are. Just as many agricultural scientists now meet with agribusiness representatives at conferences, field tours, and luncheons, progressive scientists should get together with organisations of agricultural workers so as to assess, formally and informally, their research needs. They may find themselves more often at benefit dances, union meetings, demonstrations, and on the picket line, and less often at faculty cocktail parties, or conducting impersonal interviews.

(b) To develop transitional technology. This could mean transferring devices (from minicomputers to insect pins), information (from statistical advice to agricultural journals), and progressive technicians (from computer programmers to entomologists). But it could also mean working for a science and technology by the people as well as for the people. Worker-controlled devices and methodologies for monitoring workplace hazards, say pesticides in the field, could not only enable workers to avoid the hazards, but could help them raise consciousness, organise, and demand improvements. In socialist countries, training agricultural workers in integrated pest management (IPM), for example, would be developing a science by the people, and would be transitional in the sense of reducing dependence on pesticides imported from the capitalist world.

(c) To expose unacceptable technology and to publicise transitional technology. Exposure of the faults of some technologies justifies demands for alternatives from the public. Conversely, the existence of alternatives makes undesirable technologies less tolerable. If the alternative technologies most publicised are transitional in the current social context, the dynamic of mass rejection of 'bad technologies' and demands for these transitional alternatives, can have a destabilising effect on the status quo. The communication media used should be accessible to the people the scientists want to serve – union newspapers, radio, foreign newspapers, farming magazines, and so on.

We have tried to show that technology and the social system profoundly affect each other, that technology is politically biased in its development and implementation, and that therefore in principle technology might be devised to aid exploited groups instead of exploiters (transitional technology). It is argued that since technology is not politically neutral, progressive scientists need to consciously choose sides even in their scientific work. Having outlined a programme of tasks, we now examine some instances of putting the programme to work.

Two examples of practice

The New World Agriculture Group (NWAG) is an organisation of North American scientists dedicated to progressive agricultural development.[15] Consisting mostly of ecologists and social scientists, NWAG attempts to discover and develop alternative methods of agricultural production that are ecologically rational, in the sense of protecting the environment and preserving long-term productive capacity, and that oppose the exploitation of workers and the unequal distribution of wealth within or among nations. Two of the projects in which NWAG members are currently participating exemplify its efforts to put into practice some of the ideas discussed earlier in this chapter.

First, NWAG has begun a programme of collaboration and technical assistance with the people of Nicaragua. Decapitalisation following the 1979 revolution which overthrew Somoza, and damage due to the war itself, crippled production and left Nicaragua with huge debts. Along with thousands of other 'Internacionalistas' from around the world, NWAG hopes to aid in consolidating the revolution; helping to increase agricultural production in ways that meet the needs of all the people, and that minimise dependence upon expensive imported chemical inputs such as pesticides and inorganic fertilisers.

Second, members of NWAG have long been active in support work for the Farm Labour Organising Committee (FLOC).[16] Taking inspiration from this political work, we have directed our scientific research towards methods of agricultural production that may be useful in FLOC's struggle by: (a) countering the spread of agricultural mechanisation as a means of breaking labour unions; (b) collecting and evaluating information about pesticides and farmworker health and safety in the mid-western states; and (c) developing alternatives to heavy reliance on dangerous pesticides.

Collaboration wth Nicaragua

In February 1981, an informal delegation of nine NWAG members visited Nicaragua to talk to agricultural officials and researchers about possible co-operation. In August of the same year, a second delegation of five NWAG members formally established with Nicaraguan officials a programme of collaboration between NWAG and Nicaragua. The statement of purpose of the collaboration is as follows:

> (a) To aid the Nicaraguans in their efforts to develop agriculture in a manner which is in harmony with the revolution's goals, by:
> (i) increasing yields, both in terms of food production and economic value;
> (ii) reducing the vulnerability of the agricultural system to natural disasters and economic uncertainty;
> (iii) developing a technology which protects the health of agricultural workers and the environment.
> (b) To help the scientific community in Nicaragua so as to achieve intellectual autonomy free of dependence on imperialist science, promote the integration of theoretical research with the achievement of practical goals, and encourage the kind of science which can see technical problems in their social and human context.
> (c) To express our own solidarity with the Sandinista revolution and defy any blockade which the US government may impose.

Underdevelopment in the Third World is a result of development elsewhere.[17] The economy of Nicaragua was based on export crops which enriched the few, rather than on food production and industrialisation that would meet the needs of the majority.[18] Nicaragua remains dependent on cash crops for foreign exchange to repay debts and rebuild the country, while at the same time is attempting to develop self-sufficiency in basic needs.[19] With a legacy of extreme underdevelopment, the graft and oppression of the Somoza government, and the criminal destruction that the Somocistas dealt to the country at the time of the revolution (both in terms of heavy de-capitalisation and actual sabotage of capital that could not be removed from the country), Nicaragua's underdevelopment is perhaps the worst in the western hemisphere. While the Sandinista government has made some truly remarkable gains, a great deal of work remains to be done.

The ecological rationality of the new government, at least in terms of agro-ecosystem development, is impressive in the light of the enormous economic pressures upon it. Development projects are thought of in terms of their potential for reducing expensive inputs, producing nutritious food, and so on. In short, the officials with whom we deal have a consciousness that one would simply not expect to find at a similar level in the USA or Canadian government, or in other Latin American governments with which we are familiar. The revolutionary spirit that probably comes from such a recent liberation movement seems to have spilled over to things ecological, including agriculture. A billboard in central Managua reflects that attitude: 'Our flora and fauna represent what we were fighting for in our revolution. Conserve them!'

The collaborative effort in Nicaragua has been quite effective. In co-ordination with the plant protection department of the Ministry of Agriculture we have participated in an international team study of *Dalbulus maidis*, a leafhopper vector of a corn disease; the development of an integrated pest management programme and cropping systems for tomatoes; and several other smaller projects. In co-ordination with the soil fertility department of the Ministry of Agriculture we have been engaged in studies on biological nitrogen fixation in legume crops. Finally, in conjunction with the Research and Documentation Centre of the Atlantic Lowlands, we have participated in several ecological studies of projected development schemes. Besides helping with research, NWAG carries out literature searches in the USA and sends relevant books and articles. For example, a recent extension bulletin on the peach palm was based almost entirely on materials provided by NWAG.[20]

One may ask why it is appropriate for NWAG to work outside of North America. Some have objected that Third World liberation will best be furthered by US citizens restraining their own government. But, given advantages such as extensive libraries, researchers from the developed countries are able to help countries like Nicaragua meet needs that it simply cannot meet as effectively by itself at present. Since these advantages were ultimately obtained largely at the expense of underdeveloped nations, such collaboration seems not only proper but long overdue.

The major feature of the programme, in political terms, is that Nicaragua makes the decisions. It is assumed that the Nicaraguan people know best what is needed for their country's development. NWAG's position is to fit whatever talents it can contribute into programmes that the Nicaraguans have initiated. This does not preclude responsible criticism. Given our experience with the mistakes that developed countries have made in tinkering with the environment, NWAG will try to help Nicaragua avoid repeating the same mistakes and resist the temptation to increase production in ways, however egalitarian, that will inadvertently destroy their environment and productive base. The point is that such criticism will be the suggestions of concerned comrades, not the dictates of development profiteers, and that Nicaragua has the final say.

Co-operation with the Farm Labor Organising Committee

NWAG members have been active in FLOC support groups in various cities since 1977, doing solidarity work in support of the strike and the boycott (e.g. fund raisers and publicity events). More direct involvement has included walking FLOC picket lines in the tomato fields of Ohio, participation in FLOC demonstrations and marches, and planning boycott strategy in FLOC's annual retreats. Thus we have had opportunities to talk to many farmworkers and have developed close personal friendships with FLOC members. After participating in such support work for several years, several NWAG members decided to choose areas of research in ecological agriculture that might serve the interests of farm labour. Two prominent issues have been mechanisation and pesticide use.

As noted above, a salient feature of mechanisation has been to displace migrant workers who have become 'uppity' and challenged current labour practices, even where mechanisation may be less productive and less environmentally sound than human labour. Mechanical tomato harvesters often miss a lot of the fruit, and frequently bog down in the rainy Mid-west summers, whereupon hand labour must be brought in anyway.[21] Finding feasible alternative production technologies that rely on hand labour could slow the spread of mechanisation and strengthen FLOC's position. If such alternative methods are spurned by the food companies that dictate grower policy, this could be publicised to illustrate to farmworkers and consumers alike that the true interests of the companies are in breaking farm labour unions, not in producing more and better food.

Intercropping, growing two or more crops in the same field, promises to be one such alternative. While usually associated with the tropics, intercropping was also well known in the Mid-west until the 1920s, when mechanisation made large monocultures more readily profitable to large investors.[22] It has begun to receive renewed attention in the temperate zone with the growing interest in ecologically rational alternatives to current agricultural practices.[23] NWAG realised from its own and other research that intercropping was usually more productive, less risky, and more labour-intensive than the use of monocultures of crops.[24] Because existing harvesters have been developed with one crop in mind (except in some cases in the People's Republic of China[25]), most intercrops cannot be harvested mechanically at present. Even though we can expect that mechanical harvesting methods for intercrops will eventually be developed, during the interim FLOC would have more time to organise workers.

Also, all such labour-intensive technologies, if productive, might benefit farmworker co-operatives or union farms, which may offer a progressive alternative to the large, mechanised farms that are steadily pushing smaller farms out of business. Such co-operatives have already been established in California.[26] Alternative agricultural technologies like intercropping may be one way to increase their ability to survive in a capitalist environment.

Thus NWAG began investigating intercropping systems as a transitional technology serving the cause of farm labour. Starting with the 1980 growing season, experiments were conducted at the University of Michigan with intercrops of tomatoes and other common mid-western crops. Preliminary results have indeed been promising. In small-scale experiments tomatoes intercropped with cucumbers, soybeans, or dry beans, have yielded as much as 31 per cent more than when grown separately.

This strategy does not mean that NWAG is opposed to mechanisation, to the extent that it may eliminate the need for onerous stoop labour. The farmworkers' union UFW has long asserted that 'mechanisation should benefit everyone, not just the grower'. FLOC has welcomed mechanisation as a means of reducing the number of children and older people who must work in the fields, provided that farmworkers are the first to be trained for the jobs on the mechanical harvesters, and that the breadwinners receive adequate wages to compensate for family members no longer working in the fields. The point is that human labour should be viewed as the valuable resource that it is, not as a burden. Decisions concerning trade-offs between productivity and human working conditions must be controlled by those who are most acutely affected by such decisions, the workers.

We have thus far been too slow in publicising on a popular level our results and the potential advantages of intercropping in general. We have begun to devote more effort to writing articles for local grower magazines; for example, *The Grower*, a magazine regularly read by vegetable farmers, recently published a report of our intercropping work;[27] attending and speaking at farmers' conferences and poster sessions; and planning trials on commercial land rented specifically for large-scale demonstrations.

More recently, NWAG began to work with the Farm Labor Research Project (FLRP). Organised in 1982 by

FLOC, FLRP is a research and public education effort focused on the problems of migrant farmworkers in Ohio, Indiana, and Michigan. The project has included sociological surveys of public awareness about FLOC, and more technical work directed towards studying and improving health in relation to working conditions, including the development of methods to monitor the use of pesticides, the exposure of workers to pesticides, the actual impact of pesticides on agricultural pests (if any), and the study of alternative technologies such as intercropping and biological control. All these aspects are tied in with education and community organisation within and outside of the farmworker community.

NWAG's involvement with FLRP provides an unusual opportunity to develop transitional technology, with the people, and by the people. A good example is the FLRP Pesticide Task Force. The tasks of the scientists participating in the task force were planned in meetings with FLOC organisers. The work began with the collection of information specifically about the pesticides that are used on tomatoes in Ohio and Michigan. Pamphlets and other media materials were produced to inform farmworkers about pesticides, health effects, and legal rights. While drawing on previous work from other areas for inspiration, as well as for information, efforts were focused on gathering a manageable amount of information in adequate detail to be most useful for this particular struggle.

Initial findings were presented as a reference manual[28] and a set of talks at a meeting of FLOC organisers who were preparing to go into the field to talk to workers. The manual includes a brief description of the pesticide problem, pesticide poisonings, health and exposure effects, carcinogenicity and teratogenicity. The manual concludes with a brief discussion of excessive use of pesticides and potential alternative methods of pest control. The talks were intended to facilitate organising by FLOC, not just to 'help farmworkers' with palliatives that do not also promote change of the structures that create pesticide problems in the first place. Various speakers discussed common pesticides and their physiological effects, pesticide package labels, legal rights with respect to pesticide use, the loss of effectiveness of pesticides, and some alternative methods.

The following resolution, adopted by the membership at FLOC's second constitutional convention, had been drafted by FLRP:

> Whereas, many of the pesticides used in the Mid-west are highly toxic, both in terms of acute toxicity and in terms of long-range effects, such as cancer and birth defects, and
> Whereas, farmworkers are continuously exposed to pesticides, and suffer from illness, disability and reduced life span, and
> Whereas, cases of pesticide poisoning typically go untreated, unreported and uncompensated, and
> Whereas, the use of pesticides results in environmental destruction, and
> Whereas, effectiveness of pesticides is often questionable, and can even make pest problems worse,
> Therefore be it resolved that FLOC denounces the indiscriminate and unnecessary use of pesticides in the Mid-west, and
> Further be it resolved that FLOC calls for strong regulations regarding the use of pesticides, exposure of farmworkers to pesticides and compensation in the case of pesticide poisoning,
> And be it further resolved that FLOC voices support for the development of alternative methods of pest control,
> Furthermore be it resolved that a permanent task force be developed by the Farm Labor Research Project and FLOC to study pesticide effects, to educate our members and to take action in appropriate ways on this crucial issue.

The short-term goal of this part of FLRP is to assist FLOC organisers in their struggle, but we hope that the project will also generate interest, participation, and enthusiasm in the research on the part of organisers and farmworkers, and lead to the participation of farmworkers in the collection of data and information. FLRP has begun to carry out research in the areas described here on a small farm in Ohio recently purchased by FLOC, and will help to plan the farm's development.

On a more long-term basis FLRP will attempt not just to address the problem of safe use of pesticides, but to challenge the current use, development, and research of agricultural technology. FLRP will promote methods such as intercropping and integrated pest management as alternatives to pesticides. Intercropping, as already noted, may increase overall yields by reducing the impact of crop pests. Biological control of insect pests has been shown to be technically feasible in processing tomatoes in California,[29] and *Bacillus thuringiensis*, a pathogen attacking caterpillars, is already listed for use on tomatoes and other vegetables in Ohio by the Agricultural Extension Service of Ohio State University. These examples of existing alternatives to the use of pesticides strengthen the demands by farmworkers for safer working conditions, by refuting objections that pesticides are a necessary evil of production. Another possible benefit of instituting low-chemical pest control programmes is the retention of existing employment opportunities – with vastly improved working conditions. Such programmes depend on frequently surveying the field to see whether pest numbers actually present a threat to the crop (often they do not and spraying is needless) – the job of surveying could be performed by farmworkers with only minimal training.[30]

In a collaboration such as ours, there is always the risk that scientists, even progressive ones, will tend to revert and apply academic criteria in deciding what is to be done. In the FLRP planning meetings the presence of FLOC members was often necessary to prevent the rest of the group from getting diverted into questions that were too removed from the immediate goal – organising. For example, we easily digressed into discussions of how we might gather sufficient statistical data to quantify the degrees to which pesticides were related to sickness among migrant farmworkers. FLOC, however, was more interested in training organisers to quickly recognise, document, and report specific cases of pesticide-related accidents and violations, for use in publicity and possible legal action. This illustrates the necessity of working in close alliance with the group that one wishes to serve.

Summary

In summary, in these projects we attempt to use our knowledge and skills as agricultural scientists to contribute to work that promotes progressive social change, such as strengthening the position of farm labour in the Mid-west or the independence of revolutionary Nicaragua. An indispensable feature is that we take our direction from meaningful collaboration with the organisations with which we ally, rather than attempt to impose our misconceptions upon them, as 'experts' have too often done in the past. If successful, these projects may become examples of how workers can do science, how progressive scientists can challenge current methods of doing science, and how both may challenge the course of science and technology, currently viewed as inevitable.

Out of the wasteland

The history of agricultural development provides an illuminating example of the political nature of science. The current direction of agricultural research, technology, and practice is not inevitable, but rather represents only one choice – that which corresponds to the class interests of only a small number of people. It would seem that agriculture, as other industries, can be viewed as a continual struggle between conservative and progressive forces, although the exact form of that struggle may be quite complex in particular instances. It is within the context of this struggle that technological change comes about. However, up to the present time technology has always seemed to serve those groups which seek to preserve the older forms of political organisation, whether it is pesticide companies in Nicaragua or Campbell's Soup Company in the mid-western USA. If we view the conflict of the two forces metaphorically as a war, technology becomes a weapon used in that war.

If technology is used as a weapon in political conflict, progressive scientists can turn the weapon around, so to speak, and develop technology specifically designed to aid the oppressed in their struggle. This is the germ of the concept of transitional technology. In contrast to so-called alternative or appropriate technology, this new approach does not just address the relationship between people and nature, growing organic vegetables and such, but also consciously recognises class conflict as the major driving force of history and seeks to intervene in that struggle on the side of oppressed classes. Thus its major function is to facilitate the transition to a new society.

The pursuit of transitional technology is a theoretical goal to which progressive scientists may aspire. The complexities of political change will almost inevitably make the analysis of 'what is a transitional technology' difficult. But the analysis itself is an important part of the progressive scientist's methodology. The new scientific method includes the political questions of how particular technological trajectories interface with on-going political conflict.

Among the political questions that need to be considered is the question of how technology has already evolved in the developed world and whether that same course truly facilitates development goals of the Third World. For example, an uncritical acceptance of the capitalist notion of efficiency and an excessive deference toward the achievements of technology under capitalism can carry over into socialist development, contributing to a continuation of capitalist intellectual domination there. By adopting methods of production initially formulated within a capitalist framework, many 'socialist' countries have introduced forms of social organisation and control that are essentially capitalist in nature in order to make effective use of this technology. The alternative is not to return to pre-capitalist ways of farming, or to go back to nature, or a small-is-beautiful revolution against the suffering engendered by imperialism. Development need not follow a single, inevitable pathway. New social relations make it both necessary and possible to create new criteria of 'efficiency' and new knowledge to serve it. In capitalist societies the social organisation and ideology prevent the available resources from being aimed at these goals. In developing socialist societies the advanced social relations are in part held back by lack of resources, including trained people. The real urgencies of production encourage the adoption of existing technologies. But progressive scientists in the advanced capitalist societies can use some of the intellectual resources accumulated from the plunder of the world by imperialism, combine them with their social and philosophical insights, and help produce the science which cannot be applied here nor as yet created there.[31]

Alternative research and work in agriculture, as well as other scientific disciplines are subject to a variety of pressures, both external and internal. External problems include finding sources of funding, routes of publication (both traditional scientific outlets as well as alternative ones more popular in nature or directed to other interest groups), academic acceptance (e.g. degrees, tenure), and so on. Internal problems are many, as we have come to realise through personal experience. It is difficult to overcome the conditioning and background that lead us into doing unconnected basic research rather than tying research to specific practical problems. We also have the tendency to ignore short-term political consequences of a given situation, concentrating our analysis and practice on 'loftier' goals. Furthermore, there are frequently severe problems involved in defining transitional technologies for particular situations. Finally, organisations of progressive scientists face the same challenges that other progressive organisations face – for example, how to deal with or even recognise sexism, racism, and elitism; and how to simultaneously avoid excessive division of labour and inefficiency.

A major problem, both internally and externally, is the communication between scientists and workers, workers and workers, and scientists and other scientists. Only through full communication can we reconcile the differences between short-term and long-term goals and resolve conflicts about alternative pathways of development. Scientists have to be constantly aware that someone else's development or very survival may depend on the decisions they are involved in.

This need for communication indicates an important direction for future work. We have often been frustrated and amazed by the lack of mutual awareness and support by scientists on the left. For example, FLRP needs to communicate more with similar projects, and many groups who work in solidarity with Nicaragua seem to be virtually unaware of each other. Since it is not the left that control the media, the need is all the more acute for networking and mutual support among progressive scientists – to exchange ideas and avoid needless replication of effort and mistakes, to resist outside pressures, and to help each other develop truly progressive personal politics.

Thus, we emphasise the need for progressive scientists to work in meaningful collaboration with the oppressed, and to co-operate fully with each other. Perhaps this simply reflects the fact that in order to overturn the rule of the powerful few, the rest of us must be united. Despite their longing for the quiet of the ivory tower, academic workers cannot consider themselves above the struggle.

11. 'They're worse than animals': animals and biological research

Lynda Birke

Throughout this book, the different authors have taken issue with the common idea that science is invariably progressive, contributing ultimately to human good. But as we have seen over and over again, science cannot be seen as somehow separate from the values and attitudes of the society in which it occurs; and as such, it has contributed to human oppression as well as to human welfare, and continues to do so.

However, it is not necessarily only humans for whom the values of science can be less than beneficient. This chapter deals with the welfare of animals within the science that we now have. This is a topic that is rarely addressed in critiques of science; indeed some critics might argue that the struggle against human oppression within science should be given priority and that, therefore, the question of animal suffering is unimportant. Yet there may be connections between the two: in this chapter, it is suggested that the values attached to animals within the scientific world view have parallels in the values attached to less powerful human groups. As such, the issues raised by consideration of animal rights may be significant for the development of our radical critiques of science.

One of the central concerns of animal welfare groups at the present time is the fate of the laboratory animal. The first part of this chapter looks briefly at the background to this concern, and at the *kinds* of arguments that are put both for and against the use of animals in research. The second part steps back from this current dialogue and outlines some of the history of ideas that have shaped these present views. In the final section the argument comes back to the present day, and suggests links between the issues of animal welfare and some of the other political concerns that are represented in this book. The intention is to sketch in some of the connections; it is not possible to do full justice in a short chapter to the plethora of arguments surrounding the issues of animal welfare in research. Those interested in these issues may find the references useful.

Becoming concerned

If you were to be taken back in time to the streets of the late eighteenth or early nineteenth century, you would no doubt have a number of shocks. One shock might well be the condition of the animals that were used for a variety of purposes. Dogs, cats, horses, mules, oxen – all were at times treated in ways that would not be generally acceptable today. Many could be seen in a thoroughly pitiful state, beaten, whipped and starved until they finally dropped dead on the street.

At the same time, you might see similar treatment of certain groups of humans. If, for example, you went to some of the big plantations of the American South, you would be just as likely to see human beings in a similarly piteous state, whipped and worked to death as slaves. Of course, at the time, many people – and particularly those who *owned* or controlled the animals or people – felt that it was acceptable and unproblematic to behave in these ways. There was a strong belief that each individual had a particular slot into which he or she was born, and that entitled those in higher slots to a moral superiority over those in lower ones; besides, by beating them up, you might extract more labour from them. If they dropped dead, well, there were more where they came from. Thus, ill-treatment of slaves was justified on the premise that slaves represented a lower (and numerous) form of humanity, over which the slave-owners had both property and moral rights. Similar beliefs were held about animals and their relationship to humans. But not everyone shared these views, and, as the nineteenth century progressed there was increasing effort to change them.

One of the arguments of current animal rights campaigners is that we do not have a *moral* right to treat animals substantially differently from the way that we should expect humans to be treated. Thus, if we are concerned about the oppression of particular human groups then, it is implied, we should be concerned also with the oppression of animals.

The connections between the struggle against human oppression in its many forms and those against animal oppression are not often made today. By and large, they involve quite different sets of people. This was not always the case, however, and many people involved with particular human struggles in earlier centuries saw their work as representing part of a defence of the poor and helpless in general. This was often seen as logically including the defence of animals. Many involved in the anti-slavery campaign, for example, were similarly opposed to cruelty to animals, arguing that some form of legal protection should be extended to both as a moral principle. In 1780, for example, Jeremy Bentham wrote:

> Why should the law refuse its protection to any sensitive being? The time will come when humanity will extend its mantle over everything which breathes. We have begun by attending to the condition of slaves; we shall finish by softening that of all animals which assist our labours or supply our wants.[1]

Similarly, a number of feminists of the late nineteenth century saw connections between their struggles for the emancipation of women and campaigns against cruelty; Frances Power Cobbe, for instance, an active leader of the anti-vivisection campaign, was also involved in the struggle for female suffrage.[2]

From the beginning of the nineteenth century, public opinion in the UK was beginning slowly to shift and attempts were made for the first time to introduce legislation to restrict certain forms of cruelty to animals. At first, these met with derision, so entrenched was the view that people were entitled to do what they liked with animals in the name of individual liberty. Finally, however, one such attempt succeeded, and in 1822, Richard Martin, then MP for Galway, managed to push a bill through parliament designed to prevent cruelty to several kinds of domestic animals.[3]

The most significant legislative change of the nineteenth century for laboratory animals came, however, some fifty years later. In 1875 two bills came before parliament both, to differing degrees, including requirements that animals only be used for scientific experiments if proper anaesthesia was used. Neither bill was sufficiently well worded, so some debate ensued and eventually, an amended bill was produced, which finally became the Cruelty to Animals Act 1876. It has been in force ever since.

In theory, the 1876 Act embodied several humane principles. It restricted, for example, the infliction of pain on animals during experiments without anaesthetics, and it restricted the use of paralysing agents (such as curare) which can cause considerable pain without the animal being able to do anything about it. However, it has met with much criticism of late on the grounds that it also contains loopholes, so that ways can be found round the humane restrictions.[4]

The 1876 Act was framed at a time when *relatively* few potentially painful experiments were carried out. Of course, if you are the individual animal being subjected to a painful experimental procedure it matters little whether there are five hundred others like you, or five million. On the other hand it is a different matter to exert legislative control over experiments involving a few animals, and over experiments involving many. Approximately five million 'experiments' on living animals are recorded annually in the UK according to Home Office statistics (although the definition of an experiment is somewhat ambiguous).[5] Many of these animals are subjected to only mildly painful or stressful procedures, such as a single injection, while others endure considerable and protracted pain – for example, as a result of injecting toxic substances into them.

Why, in a country which prides itself on being a 'nation of animal lovers', is such suffering tolerated? Mainly, because most people consider it to be ultimately beneficial for the progress of medical research. The suffering of a few animals now might be worth while, if, in the end, medical science can produce, say, a cure for cancer. Moreover, the end-product may also benefit those animals we keep domestically. There are sometimes outcries that *particular* experiments are cruel to animals not because these are necessarily more painful, but because they are seen – and portrayed by the press – as being of *less* benefit than many others.[6] The overriding assumption then, is that such exploitation of animals is justified on the grounds of some putative medical benefit.

Now there are many problems with such a position. How, for instance, do we arrive at the assumption that it *is* morally justifiable to ill-treat and sacrifice some animals for a putative greater good? If we are to balance one against the other, how do we decide where the balancing point is to be? Is the 'medical benefit' one on which we would all agree, or are some kinds of medical knowledge more justifiable than others? Why can we make this assumption about animals, but not about other humans? The position of belief in a greater good is apparently not a straightforward one.

A second, and significant, problem with the commonplace view is that it ignores the political and social context of the science/medicine in whose name the animals are suffering. As many other chapters of this book have indicated, this science is certainly not always acting for the greater good of humanity, let alone other animals. An assumption behind the 'medical progress' justification is that the driving force behind all decisions to use animals in obviously painful procedures has been a decision that the procedures are for the ultimate human good. In practice, of course, such decisions are guided far more strongly by financial considerations. It is not medical progress (howsoever defined) that decides, but money. Harrison[7] notes the kind of double standard that is entailed:

> If one person is unkind to an animal it is considered to be cruelty, but where a lot of people are unkind to a lot of animals, especially in the name of commerce, the cruelty is condoned, and, once large sums of money are at stake, will be defended to the last by otherwise intelligent people.

It is indeed the use of animals by obvious commercial interests that often arouses the most ire on the part of animal welfare campaigners. The cosmetics industry in particular has come in for heavy criticism precisely because it uses large numbers of animals, in often rather painful tests, to test a range of potential cosmetics. Thus, many animals are made to suffer adverse, painful skin reactions in order to provide something which can hardly be said to be for the greater good of humanity.

On the one side are those who would defend the use of animals, whose chief line of defence seems to be to point to the medical benefits and argue that these *could* not and would not have occurred had we not used animals. On the other side, are those opposed implacably to any kind of experiment using animals in the belief that it can never be morally justified. In between are a wide range of opinions. Although public opinion on the issue appears to be shifting somewhat, there is not likely to be a major change in the legal position in the near future despite lengthy debates both in the UK and within the European Community.[8] One important reason for this stasis is that public opinion is remarkably ambivalent on the issues involved. For instance, while often opposed to a wide range of experimental uses of animals, the same people may feel equally strongly that whatever chemical substances they deal with at home or at work should have been adequately tested for safety. And generally, such testing involves animal experiments.

Our attitudes concerning the use of animals in the name of scientific research are, then, often deeply contradictory. There are other contradictions that are worth noting: for example, there is the concern for the fate and suffering of at least those creatures with whom we can have some empathy such as mammals, or at least those mammals that are not regarded as pests; hence, there is more concern expressed for cats than rats. This carries with it the implicit belief that such animals have at least *some* claim to certain rights. But in opposition to this is the strand of belief that maintains that *we* have a right to use animals for our own ends, that it is always permissible to have dominion over them, and exploit them. These two beliefs cannot easily be reconciled.

To a degree, of course, this attitude of exploitation towards other species has existed for as long as human societies have existed; humans have always exploited other animals for food, for their skins, and so on. What is significant to note, however, is that this belief has *intensified* considerably over the last four hundred years. As the scientific revolution progressed, hand in hand with the rise of capitalism, so the exploitation of animals grew. Animals were, after all, part of the natural world that was rapidly being plundered for greater profit.

'Nature' and the rise of science

A distinction that is now central to our view of nature, including other species, is that between 'nature' and 'culture'. Over the last hundred years or so, this dichotomy has become deeply embedded in Western thought and in the theories produced by science about nature.[9] We have come to view 'nature' as somehow rather chaotic and disorderly, whereas by contrast, 'culture' is seen as the epitome of progress, implying human mastery *over* nature. Science today is an intimate part of that distinction, for it is science that has promised to give us technological mastery over our environment.

The nature–culture dichotomy is so prevalent in our ways of thought that we tend to take it for granted without seriously beginning to challenge it. We tend to conceive of nature as being 'out there', something which we can study in minute detail, but not something of which we are an integral part. There can be no introspection in our study of nature: we are merely the observers and measurers.

In part, this dichotomy has some of its origins in the ancient, pre-Darwinian idea of a 'Chain of Being' (see chapter 9), which was the dominant theological doctrine for many centuries. According to this, creation was arranged in a linear hierarchy, with God at the top, then humans (with, it should be noted, men above women and Europeans above other races), and then, arranged equally linearly below, came all the other species of animals and plants then known.[10] The conception of this chain of being implied a hierarchy, and it was one in which humans and human culture were held to be superior to other forms of life. Those nearer the top of the scale were by definition closer to God, more likely to possess a soul. European man, for instance, without question had a soul; there was rather more debate concerning the possibilities of souls for other human races, women, and animals. Those further down, were 'lower' forms of life. To say of someone that 'they are worse than animals' is to imply that they, too, are a lower form.

However, while this idea was important, it is also important to note that for much of its history, it co-existed with other ideas that placed much less stress on hierarchies of creation. One such set of beliefs was that the cosmos represented not a linear hierarchy but an organic unity. Within this, the Earth was seen as a nurturing mother that was sensitive and alive, and capable of responding directly to human action; she was the matrix that gave birth to all living things, whether human or non-human.[11] This was, then, an essentially holistic view of the living world, one which emphasised the common spirit pervading all organisms and which stressed their interrelatedness. To many of its adherents, other species of animals had certain rights and they should be respected for their sentience and sensitivity just as we would respect the sensitivity of other human beings.

Slowly, this holistic, organic view of nature began to give way to a more mechanistic conception. The rapidly expanding market economy of the early sixteenth century helped to create a new relationship with the Earth and nature: mining for minerals, for example, increased rapidly to meet expanding market needs, and crops were no longer grown principally for local consumption, but were grown for profit. For a time the two images of nature co-existed, but in the end, the holistic view lost out to the more profitable mechanistic world view that has come to characterise science. Not only was the latter actively encouraged by the emerging bourgeoisie, since it provided legitimation of their exploitation of natural and human resources, but also there was considerable repression of those who held alternative views.[12]

Views of nature which emphasised co-operation and harmony came up against more and more challenges from the mechanistic world view. Organisms and ecosystems became increasingly viewed as machines, whose inner workings could be understood in terms of mechanical analogies. To Descartes, for example, the human body operated according to mechanical laws, just as did clocks or other pieces of machinery.

And what of the fate of animals in this changing conception? To begin with, they were *part* of nature, over which human culture was rapidly gaining more and more control. In the rigid distinction between (inferior) nature and (superior) culture that was emerging, animals came increasingly to represent part of exploitable nature. As noted above, as long as there were prevalent ideas of the harmony of the Earth, there existed some kind of constraint on the extent to which humans could exploit nature. As the scientific revolution progressed, however, the view of nature as dead inert matter – mere mechanism – gained ascendancy over the view that humans might co-exist peacefully and harmoniously with other forms of life. For animals, this meant that, being part of nature, they were consigned to the category of machines. Descartes elaborated on this: for him, the body – whether animal or human – *was* simply machinery. What distinguished humanity from the rest of brute creation was the possession of a soul. This dualism did not allow for sentimental feelings towards animals: a woman disciple of Descartes, Mme de Grignan, objected most strongly to the suggestion that a dog be given to her daughter: 'We want only rational creatures here, and belonging to the sect we belong to we refuse to burden ourselves with these machines.'[13]

This mechanistic view of animals became absorbed into the values of the emerging sciences and influenced the ways in which animals were used in the laboratories. During the eighteenth and nineteenth centuries, for example, many physiological experiments were conducted on living animals, long before the days of anaesthesia, thus inflicting terrible pain. Even the screams and howls of the tortured animals were assimilated to the notion of mechanism: the howls were merely the grinding of machinery. Even with the advent of clinical anaesthesia, little changed in the physiology laboratories: dogs were still being nailed to boards and eviscerated while still alive; and horses were still being subjected to endlessly repeated surgical operations without anaesthesia to give medical students 'practice'.[14]

It is precisely this view of the expendability of animals in the laboratory to which we are heir. Animals in the laboratory lost their individuality, with feelings and experiences of their own, and became ciphers, mere numbers in a statistical game. If one were to die by accident during the course of some experimental procedure, the scientist is not expected to express emotion, but must retain a cold distance; there are, after all, more where that one came from.

Individual scientists accept this kind of philosophy to varying degrees. While it is undoubtedly true that some biologists would eschew methods that are liable to inflict considerable suffering, it is also true that there are many who do not; many who are so desensitised to the suffering of the animals that they routinely use that they no longer see it. Although the law (in the UK at least) requires the use of anaesthesia, there are many kinds of experiment in which anaesthesia 'would frustrate the object of the experiment'; and even when it does not, anaesthesia does not usually last for long. And there has yet to be a scientific report which describes methods of post-operative analgesia.

These types of attitudes towards animals are not necessarily unique to science, of course, and animal welfare groups have made much of similar attitudes in, for example, modern intensive farming. The salient point, however, is that these practices have in common a view of animals as machines, as expendable commodities – a view that has been strongly influenced by the now dominant world view that we associate with science.

The rights of animals

For many people, this apparently callous disregard for the sentience and feelings of the individual animals used in laboratories is indeed the hallmark of modern science: for those who hold such views, science should not be allowed to continue in this way without challenge, but should proceed only with the use of alternative, non-invasive methods.[15] Whatever the feasibility of seeking for alternatives, the central focus of this opposition is that the question of whether or not to use animals in research is a *moral* one, including the principle that animals have *rights* that we should respect. Hence, for adherents of this view, it is irrelevant whether or not we currently have the expertise to develop a wide enough range of suitable alternatives to satisfy all research needs: *any* experiments which use animals in the meantime are considered immoral.

The question of rights is not necessarily straightforward. If you extend the concept of 'equal rights' to certain animals, then you are merely moving the boundary outward from humans; and where do we then draw the line? Do we extend the concept only to vertebrates, or do we extend it to cover, say, earthworms and bluebottles? Or is the concept of equal rights only applicable to those species of animal in which we can easily *identify* suffering, because they are built approximately like us? Evidently, it is not going to be an easy concept to define. However, that it may be difficult to define is not necessarily important; it may be of greater importance first to establish that there is a case for considering the rights of at least some kinds of animal.

This is where the trouble really begins. The dispute over the use of animals in experiments is not, as is sometimes implied, a dispute *over* a moral issue, but has more to do with whether there *is* a moral issue in the first place. In discussing this, Cora Diamond[16] outlines the two opposed viewpoints regarding the use of animals in experiments. First, there is the view that defends the use of animals:

> Within certain limits experimental animals may be regarded as delicate instruments, or as analogous to them, and are to be used efficiently and cared for properly, but not more than that is demanded.

This is counterposed to the second type of view:

> Within certain limits, animals may be regarded as sources of moral claims. These claims arise from their capacity for an independent life, or perhaps from their sentience, but in either case the moral position of animals is seen as having analogies with that of human beings.

In other words, those who would justify the continuing use of animals in research do not see it explicitly as a *moral* issue, while for those opposed it is precisely the ethics of using animals in research that is questionable. Underlying these distinctions is the question of the extent to which we treat human beings as special. Thus, using animals in research might be justified on the basis that humans are special in some kind of way, and that this *entitles* us to use other species for our own ends. By contrast, the opponents of animal experimentation might argue that an individual can be the subject of moral concern if, for example, it possesses sentience, or the capacity to lead an independent life.[17]

Neither position is completely adequate. It is quite possible to believe that humans do have certain special qualities (such as the richness and complexity of human social organisation) and at the same time to see the similarities and continuity between ourselves and other living organisms. It is somewhat paradoxical that scientists defend an essentially discontinuous position with respect to the ethics of experimentation, in which humans are far removed from other species. Thus, for example, *no* experimentation on humans can be permitted, whatever their sentience or intelligence, while any experimentation on animals, whatever their sentience or intelligence is in theory permissible. Yet at the same time, they usually justify research in terms of the relevance of animal research to humans, thus implying that there are similarities after all.

To this author it is not completely obvious that we are justified in always distinguishing between 'special' humans, for whom we should show some kind of species loyalty and who alone can be the subjects of our moral concern, and other animals, who can have no such claims. Appeals to the specialness of human beings do not seem to justify some of the things that are done to animals in laboratories. Appeals to a spurious species loyalty cannot justify the use of animals to test items such as cosmetics; and the use of animals in what is euphemistically called defence research is quite insupportable. Subjecting animals to high doses of radiation to bring on radiation sickness cannot, surely, ever be justified – any more than it can be when it is done on humans.

Although it should be evident from this brief sketch that the issue of animal rights – and whether or not the questions raised are essentially *moral* ones – has no easy solution, it is becoming increasingly clear that it is an issue with which scientists must be concerned. Whatever their position on the issue, scientists have now to face a greater awareness – and even opposition – than they have done for many years. Vigilante groups calling for 'animal liberation' are breaking into laboratories, and the press continue to write lurid accounts of what happens to dogs, cats, or other cuddly animals in laboratories. There are more and more calls to make scientists more accountable to the public in the way that they use animals. Animal rights are rapidly becoming a topical issue in relation to the progress of science. How, then, might we try to draw connections between this issue and other facets of the critique of science with which this book has been concerned?

Animals and the political critique of science

There are two underlying questions to be raised in this section: first, is it relevant to discuss our treatment of animals in connection with our critiques of science; and second, if it is, what sort of consideration should we give to animals in our discussions of a new kind of science?

Many of the radical critiques of science to date have made the assumption that, given appropriate social change, it would be possible to conceive of a more humane and radical kind of science. Such a science would, for example, be geared more to the needs of people than to the needs of profit. As long as we do not raise the questions of methodology, this vision seems highly laudable; when such questions are raised, however, there is likely to be a conflict of interests particularly if those methods include animal experimentation. Presumably our more humane science will still involve, for example, the manufacture of a range of chemicals for various uses and we would clearly want these chemicals to be adequately safety-tested.

But then the problem arises: *how* do we carry out safety tests? At present, the criteria used include the LD_{50} test, which calculates the dose of the chemical required to kill 50 per cent of animals tested. The remaining 50 per cent, of course, may not die, but may suffer considerable pain from exposure to the substance being tested. A variety of other tests may be used in addition, depending on the nature of the chemical under test. What is important, however, is that, *at present*, routine safety testing of drugs or other chemicals is likely to involve exposing animals to toxic or lethal levels of the compound.[18]

So if we are to develop a more humane science we need, first, to develop adequate methods of safety testing. This certainly means better methods and higher standards than are currently used. But at present, given that we have not developed alternative, non-invasive methods of testing substances, we appear to have no other way of doing this than by using animals in a way that is, for many people, exactly the antithesis of humane.

Such use of animals for safety testing may seem necessary in terms of the health and safety of those humans who have to live or work with the chemicals involved. It does, however, rest on the distinction that the welfare of humans is *always* of greater moral significance, and also that animals exist as objects for our use. But if the idea of a more humane science is to be pursued, why should we only be humane to our own kind? Why do we not extend that principle to animals?

Creating a science with more liberatory potential, a more humane science, may mean different things to different people. One aspect includes not only the development of a science more explicitly geared to human need, but also the development of a more holistic, non-reductionist science. The reductionist philosophy, outlined in chapter 1, has underwritten modern science, and has contributed greatly towards the conceptualisation of the universe in terms of atomised bits that might be understood by mechanical analogies. The mechanistic world view has taught us to separate 'nature' from human 'culture', and science teaches us how to understand 'nature' by tearing it apart into smaller and smaller bits.

Carolyn Merchant, in her excellent book *The Death of Nature* describes how the emerging mechanistic philosophy carried with it certain consequences for our views of nature:

> Not only did this new image function as a sanction, but the new conceptual framework of the scientific revolution – mechanism– carried with it norms quite different from the norms of organicism. The new mechanical order and its associated values of power and control would mandate the death of nature.[19]

Those values of power and control have, as we know only too well, brought us much environmental destruction. Yet, although the death knell may have sounded, there remain some traces of more holistic views of nature. Merchant suggests that one important source today is provided by the ecology movement, with its emphasis on co-operation with, rather than mastery over, nature. In seeking a more liberatory science, this emphasis on co-operation *with* nature is surely essential. And to co-operate with that which we study requires that we distance ourselves from it less, that we see ourselves as part of it, and that we place far less stress on the mythical idea of 'objectivity'. Indeed, we *can* only be 'objective' if we do pretend that we are removed from the subject of study.

For many feminist writers, it has been precisely that undue, and unrealistic, stress on an unattainable objectivity that has contributed to the inhumane character that science now has. Women, it has been argued, have been socially assigned to the caring, subjective sphere[20] – and it is this that is sadly lacking in the inhumane, and inhuman, face of modern science. Rita Arditti, in writing about the possibilities of creating a more liberatory, and specifically a more feminist, science notes:

> Today, in science we know 'more and more' about 'less and less'. Science as an instrument of wealth and power has become obsessed with the discovery of facts and the development of technologies. The emphasis on the analytical method as the only way of knowing has led to a mechanistic view of nature and human beings . . . The task that seems of primary importance – for women and men – is to convert science from what it is today, a social institution with a conservative function and a defensive stand, into a liberating and healthy activity. Science needs a soul which would show respect and love for its subjects of study and would stress harmony and communication with the rest of the universe.[21]

If this more liberatory science is to develop with a respect and love for its subjects of study, then it is going to have to shrug off its dualistic faith in nature *versus* culture. Amongst other things, this will involve attempting to develop a different perspective on why and how we use animals in our science. That science may seek non-reductionist ways of understanding the world, but in doing so it will need to accept that animals, that are *part* of the world, may have claims to moral rights in the same way as humans do. It would hardly be any kind of 'liberatory' science that was centred around an ideology opposed to oppression and yet maintained the present mechanistic view that nature is there for our exploitation. If we are to survive, then we are going to have to adopt – with considerable urgency – a more holistic and co-operative approach to nature.

In moving towards a more co-operative view, we have also to begin to see ourselves, along with other species that make up the various ecosystems, as part of nature. We may retain the power to alter our environment in a way that no other species has yet been able, but if we are to exercise that power, we will need to do it in a way that preserves the integrity of the whole. And if we are a part of nature, existing in continuity with it, then we no longer have the automatic right to dispense with ethics and to treat other animals as expendable machines. That fixed and rigid distinction between humans and other animals, that allows us to do so much, would have far less currency if we were to live in co-operation with the rest of nature. And co-operation, not exploitation, is after all, the hallmark of a truly liberatory science.

Conclusions

The chapters of this book have covered many topics, but we hope one thing is clear: biology, like other sciences, cannot be seen as value-free and 'above' politics. Time and again, the authors of different chapters have pointed to the close connections between the theory and practice of the life sciences and the social and political context in which it takes place. The values that are sustained by the practice of the life sciences are clearly not ones that are, by and large, shared by the contributors to this book.

It is one thing to criticise the practice of existing science – to say that it supports particular ideologies and not others, that it contributes to the oppression of some groups of people, and so on – but what can be done about it? Science in the late twentieth century is massively institutionalised; it is carried out and supported by corporate interests and the state rather than by individual scientists. In the face of such a monolith, it seems impossible to change. But that said, there are attempts to make changes and it seems crucial – after the series of critiques in this book – to end by indicating some of these. By so doing, we hope we are ending on an optimistic note.

Attempts to change some features of biological practice or theory may seem an interesting, but futile exercise; futile, in the sense that, if it is accepted that science is an intimate part of a capitalist social and political structure, then it cannot be changed in the absence of changing capitalism itself. It is, of course, true that science would change if we were to change the organisation of society and its production; but equally, it seems unlikely that that will happen tomorrow, at least in the UK. Are we to sit back and accept, or simply avoid, the science that we now have until some more radical change takes place?

It is the belief of the contributors to this book that some change is possible – indeed, desirable – within science even in the absence of any more dramatic social events. Science can be changed both in its practice (e.g. in the social organisation of individual laboratories or work-places) and in its theory (e.g. in attempts to develop less reductionist theories of the life sciences). Both can contribute toward the development of a more progressive biology, hopefully one that lends itself less readily to reactionary ideologies.

Within the radical literature, there are often references made to the need to create a more 'socialist' or a more 'feminist' science. What is meant by this? In the first place, both terms are used to describe a science geared to human *needs* rather than to the profit of large corporations. Second, it would not be a science that reduced nature to little bits in order to dominate or extract from it, but would attempt to understand the natural world in terms of its complexities. Individual organisms, cells, communities – or whatever the units with which we work – would not be understood solely in terms of their component parts, but more in terms of the complex relationships between these units and others. Third, it would also be, as many feminist critics have pointed out, a science that is not preoccupied with pursuing the elusive ideal of detached objectivity, but would be one that openly acknowledged the importance of subjectivity and intuition in the process of producing knowledge.

What kind of changes might be made to our sciences? Here, we will briefly consider three: changing scientific practice; changing ideas within the life sciences (e.g. in the way that we study it); and changing the ways that we view nature and our relationship with it.

Changing scientific practice

First, as the authors of chapter 10 noted, scientific practice may be changed to create a more egalitarian and democratic organisation of work within science. In that chapter, they noted the ways in which this has been done with respect to their own scientific work. There have been a number of other attempts to democratise scientific work.[1] What is important to note, however, is that these attempts do indicate that it *is* possible to make the work process within the laboratory more egalitarian, even if the world outside it is not.

Apart from attempts to change individual laboratories, another method of altering scientific practice is to place scientific expertise in the hands of the community rather than an elite within large industries. This is the principle behind the 'science shops' that can be found in the Netherlands. These are organisations to which members of the community may go for scientific information or expertise. Their nearest equivalent in the UK is the work hazards groups set up in various parts of the country by trade unionists concerned with health and safety at work, often in conjunction with radical scientists. These groups are usually affiliated to the British Society for Social Responsibility in Science (BSSRS). Their chief role is to act as a centre, much along the lines of the Dutch science shops, for information exchange and data collection on issues of health and safety in local industries.

Changing scientific ideas

Apart from changing scientific practice, we might also attempt to change the ideas that dominate thinking in the life sciences. Two attempts in this direction might be noted here. The first is the attempt to move away from reductionist life sciences in their various guises, and towards the creation of more dialectical life sciences, in which the organism (or other biological unit) is seen in terms of a dialectical relationship with its environment. This approach allows for both the complexity of interrelationships between events, and also for the possibility that 'biology' may itself be changed as a function of that unit's history and environmental context. When the unit is individual animals, including humans, then the idea that the individual's biology may itself be a product of its history and context is an important one with which to counteract the prevalent notion that we are as we are because it is all fixed in our genes. Various approaches to this pursuit of a more progressive biology have been made in different fields of the life sciences (e.g. the Dialectics of Biology group).[2]

Another move towards the creation of more progressive life sciences lies in the attempt to emphasise a 'biology of co-operation'. Since Darwin, the notion of competition in evolutionary theory has often been used to support the ideas of competition within capitalism – a tendency which has increased dramatically during the recession years of the early 1980s. The notion of competition is increasingly used to buttress the ideology and social policies of the new right, as well as being used as a rationalisation of the arms race. Yet competition is not the only mechanism in evolution, and is certainly not the sole driving force in nature that some right-wing ideologues suppose it to be. Alongside the emphasis on competition there has been a current of ideas which emphasise the extent of *co-operation* in nature.[3] It seems important to maintain this emphasis, and to stress that co-operation in nature is widespread. By doing so, we might hope to counter the claims made by the right that socialism is a delusion that can never hope to overcome humankind's base competitive instincts. Attempting to understand the conditions in which humans *are* co-operative and nurturant, rather than competitive and exploitative – whether these involve biology or not – seems an urgent task to undertake. Co-operation and tolerance are, after all, more likely to be the cornerstones of peace.

The scientific view of nature

Finally, the ways that we view nature in general need to be changed. Capitalist society has accepted a mechanistic view of nature, in which the natural environment becomes something to be dominated, in order to further human mastery over the world. This exploitative vision has resulted in, among other things, the despoliation of the environment, a view that other living things do not really matter, and, above all, the danger of nature's total destruction in a nuclear holocaust. Whatever else we do, we have – urgently – to change the ways in which nature is understood. 'Nature' is something we have to learn to co-operate with rather than compete against if we are to avert our annihilation. Changing the ideas, and the social practice, within the life sciences – to emphasise co-operation rather than competition – seems a very good place to start.

Notes and References

Introduction

1. E. Yoxen (1983) *Gene Business: Who Should Control Biotechnology*?, London: Pan Books.
2. P. Kitcher (1983) *Abusing Science: The Case Against Creationism*, Milton Keynes: Open University Press.

1. Biological reductionism: its root and social functions

1. Part of the discussion in this chapter is drawn from Rose, S.P.R., Lewontin, R., and Kamin L. (1984) *Not in Our Genes*, Penguin, which discusses these theories in greater detail.
2. Wilson E.O. (1975) *Sociobiology: The New Synthesis*, Cambridge, Mass.: Belknap Press.
3. Dawkins R. (1976) *The Selfish Gene*, Oxford University Press.
4. Rose, S.P.R. (ed.) (1982) *Against Biological Determinism*, London: Allison and Busby.
5. Rose, S.P.R. (ed.) (1982) *Towards a Liberatory Biology*, London: Allison and Busby.
6. Lewontin *et al.*, note 1.

2. Neuroscience: The cutting edge of biology?

1. *Nature*, (1981) vol. 293, pp. 515–34.
2. Sperry R.W. (1981) 'Changing priorities', *Ann. Rev. Neuroscience*, vol. 4, pp. 1–15.
3. *Ibid.*
4. Matthews E. (1983) *Marx 100 Years On*, London: Lawrence and Wishart.
5. Mark, V.H., Sweet, W.H. Ervin, F.R. (1967) 'Role of brain disease in riots and urban violence', *J. Am. Med. Assoc.* vol. 201, p. 895.
6. Mark, V.H. and Ervin, F.R. (1971) *Violence and the Brain*, Harper and Row.

7. Chorover, S.H. (1980) 'Violence: a localisable problem?' in Valenstein, E. (ed.) *The Psychosurgery Debate*, Freeman.

8. Chorover, S.H. (1980) 'The psychosurgery evaluation studies and their impact on the commission's report', in Valenstein, E. (ed.) *The Psychosurgery Debate*, Freeman.

9. Kesey, K. (1962) *One Flew Over the Cuckoo's Nest*, Viking Press.

10. Lennard, H., and Cooperstock, R. (1980) 'The social context and functions of tranquilliser prescribing', in Mapes, R. (ed.) *Prescribing Practice and Drug Usage*, London: Croom Helm.

11. Medical Research Council (UK), *Annual Report, 1982–83*.

12. Dyer, K.F. (1984) '*Genetic determination of spatial ability: conjectures and refutations*', unpublished manuscript.

13. Money, J. (1970) 'The therapeutic use of androgen-depleting hormone', *J. Sex Res.* vol. 6, pp. 165–72.

3. The determined victim: women, hormones and biological determinism

1. See, for example, the controversy surrounding the use of the injectable hormone Depo-Provera in the UK. For example, Rakusen, J. (1981) 'Depo-Provera: the extent of the problem. A case study in the politics of birth control', in Roberts H. (ed.) *Women, Health and Reproduction*, London: Routledge & Kegan Paul.

2. Borrell, M. (1976) 'Organotherapy, British physiology and the discovery of the internal secretions', *J. History Biology*, vol. 9, pp. 235–68.

3. Walsh, V. (1980) 'Contraception. The growth of a technology', in Brighton Women and Science Group (eds) *Alice Through the Microscope: The Power of Science Over Women's Lives*, London: Virago.

4. Aberle, S.E. and Corner, G.W. (1953) *Twenty-five years of Sex Research: History of the National Research Council Committee for Research in Problems of Sex*, London: Saunders.

5. Cited by Haraway, D. (1978)'Animal sociology and a natural economy of the body politic. Part I: A political physiology of dominance', *Signs: Journal of Women in Culture and Society*, vol. 4, no. 1, pp. 29–30.

6. For instance, at the time of writing, there has been controversy over some of the findings from animal studies of Depo-Provera, since it is reputed to have caused mammary tumours in dogs. On the one hand, there are advocates of Depo-Provera claiming that these data are dubious since the strain of dog used tend easily to get mammary tumours anyway, and on the other hand, there are

those for whom any adverse findings suggest that the drug is dangerous.

7. Androgens and oestrogens are two types of steroid hormones, the other major type involved in sex and reproduction being the progestins. All three kinds are produced by both sexes, although in differing amounts.

8. See Birke L.I.A. (1981) 'Is homosexuality hormonally determined?' *J. Homosexuality* vol. 6, no. 4, pp. 35–50.

9. In connection with this, it is important to note that, in the biological/medical literature, homosexuality is frequently confused with problems of gender identity or role. Thus, since women are socially expected to adopt a feminine role, an apparently 'feminine' woman is not likely to be thought to be lesbian. By contrast, lesbianism is supposedly associated with greater than average levels of 'masculinity' (however defined). The opposite is true for males: that is, gay men are supposed to be effeminate. The hormonal bases that are then sought are in accord with this: thus, in gay men, an excess of 'female hormones' may be sought, while in lesbians, an excess of 'male hormones' is looked for. I have discussed the conceptual errors on which this is based further in Birke, L. (1982) 'From sin to sickness: hormonal theories of lesbianism', in Hubbard, R., Henifin, M–S. and Fried B. (eds.) *Biological Woman–The Convenient Myth*, Boston, Mass.: Schenkman.

10. That is, there is no evidence to link either homosexuality or transsexualism with any biological causal factor.

11. Bell, A.P. and Weinberg, M.S. (1978) *Homosexualities*, New York: Simon and Schuster.

12. Birke, note 8.

13. Holliday, L. (1979) *The Violent Sex: Male Psychobiology and the Evolution of Consciousness*, Guerneville, California: Bluestocking Books.

14. Goldberg, S. (1974) *The Inevitability of Patriarchy*, London: Temple Smith.

15. Pizzey, E. and Shapiro, J. (1981) 'Choosing a violent relationship', *New Society*, 23 April.

16. Pizzey, E. and Shapiro, J. (1982) *Prone to Violence*, Feltham, Middlesex: Hamlyn.

17. *Ibid.* p. 181.

18. Cooper, W. (1975) *No Change*, London: Hutchinson.

19. Grossman, M. and Bart, P. (1979) 'Taking the men out of menopause', in Hubbard *et al.* (eds) note 9.

20. Weideger, P. (1979) *Female Cycles*, London: Women's Press.

21. Birke, L. with Best, S. (1982) 'Changing minds: women, biology and the menstrual cycle', in Hubbard *et al.* (eds) note 9.

22. Cooper, note 18.

4. Pharmacology: why drug prescription is on the increase

1. Turner, J.S. (1976) *The Chemical Feast*, Harmondsworth: Penguin, pp. 49–66.

2. McGeer, E.G., Olney, J.W. and McGeer, P.L. (1978) *Kainic Acid as a Tool in Neurobiology*, New York: Raven Press, pp. 95–121; Rogers, L.J. (in press) 'Teratological effects of glutamate on behaviour', *Food Technology in Australia*.

3. van den Bosch, R. (1978) *The Pesticide Conspiracy*, New York: Doubleday.

4. Lee, P.R. (1980) 'America as an over-medicated society', in Lasagna, L. (ed.) *Controversies in Therapeutics*, Philadelphia, PA: Saunders; Christensen, D.B. and Bush, P.J. (1981) 'Drug prescribing: patterns, problems and proposals,' *Soc. Sci. Med.* vol. 15A, no. 3, pp 343–55.

5. Phillipson, J.D. (1979) 'Natural products as a basis for new drugs', *Trends in Pharmacological Sciences*, vol. 1 no. 2, pp. 36–38.

6. In this case, a placebo effect means that the person receiving the treatment is cured not by any direct biological action of the medicine but rather by psychological suggestion based on belief in the healing powers of the medicine. For further discussion of folk medical practices, see Camp, J. (1973) *Magic, Myth and Medicine*, London: Priory Press.

7. Weiner, N. (1980) 'Atropine, scopolamine, and related antimuscarinic drugs', in Gilman A.G. and Goodman L.S. (eds) *The Pharmacological Basis of Therapeutics*, 6th edn, New York: Macmillan, pp. 121–202.

8. Diabetics have hyperglycaemia, abnormally high blood sugar levels, due to underutilisation and overproduction of glucose. Hypoglycaemic agents lower blood glucose levels by restoring glucose metabolism to normal, in the case of tolbutamide by stimulating the islet tissue of the pancreas to secrete insulin.

9. O'Dea, K., Spargo, R.M. and Akerman, K. (1980) 'The effect of transition from traditional to urban life-style on the insulin secretory response of Australian aborigines', *Diabetes Care*, vol. 3, no. 1, pp. 31–37.

10. Skegg, D.C.G., Doll, R., Perry, J. (1977) 'Use of medicines in general practice', *British Medical Journal*, June, pp. 1561–63.

11. Study cited by Lader, M. (1978) 'Benzodiazepines–the opium of the masses?', *Neuroscience*, vol. 3, pp. 159–65.

12. Cooperstock, R. (1978) 'Sex differences in psychotropic drug use', *Soc. Sci. and Med.*, vol. 12B, pp. 179–86.

13. Owen, A. (1976) 'The psychotropic drugs: part II', in Diesendorf, M. (ed.) *The Magic Bullet*, NSW, Australia: Southwood Press, pp. 44–48. Also see Lader, note 11 and Rowe, I.L., (1973) 'Prescription of psychotropic drugs by general practitioners', *The Medical Journal of Australia*, vol. 1, pp. 589–93; and Whitlock, F.A. (1980) *Drugs, Drinking and Recreational Drug Use in Australia*, Australia: Cassell, pp. 54–57.

14. Travers, R. (1981) *Australian Mandarin; Life and Times of Quong Tart*, Australia: Kanga Press; and Lonie, J. (1979) 'A social history of drug control in Australia', research paper 8, in O'Brien D. (ed.) *Royal Commission into the Non-Medical Use of Drugs, South Australia*, Adelaide: Gillingham Printers, pp. 1–21.

15. See for example, Costa, E. (1979) 'The role of gamma-aminobutyric acid in the action of 1,4-benzodiazepines', *Trends in Pharmacological Sciences*, vol. 1, no. 2, pp. 41–44.

16. See for example, Cooperstock, note 12.

17. Pflanz, M., Basher, H.D. and Schwoon, D. (1977) 'Use of tranquillizing drugs by a middle-aged population in a West German city', *Journal of Health and Social Behaviour* vol. 18, pp. 194–205.

18. For example, Cooperstock, note 12; Skegg *et al.* note 10; Reynolds, I., Harnas, J., Gallagher, H., and Bryden, D. (1976) 'Drinking and drug taking patterns of 8,516 adults in Sydney', *The Medical Journal of Australia*, vol. 2; and Bergman, U., Dahlstrom, M., Gunnarsson, C. and Westerholm, B. (1979) 'Why are psychotropic drugs prescribed to out-patients? *European Journal of Clinical Pharmacology*, vol. 15, pp. 249–56.

19. Cooperstock, note 12.

20. Cooperstock, R. (1976) 'Psychotropic drug use among women', *CMA Journal*, vol. 115, pp. 760–64.

21. Dunnell, K. and Cartwright, A. (1972) *Medicine Takers, Prescribers and Hoarders*, London: Routledge and Kegan Paul.

22. The word hysteria comes from the Greek *hystera* meaning womb. Hysteria was characterised by fainting, convulsions, and so on, accompanied by disturbed moral and intellectual faculties, thought to stem from malfunction of the womb.

23. See, Cooperstock, note 12, and Rosenfield, S. (1980) 'Sex differences in depression: do women have higher rates?' *Journal of Health and Social Behaviour*, vol. 21, pp. 33–42.

24. For example, Skegg *et al.* note 10.
25. Bliss, M.R. (1981) 'Prescribing for the elderly', *British Medical Journal*, vol. 283, pp. 203–06.
26. *Ibid.*
27. Parry, H.J., Batter, M.B., Mellinger, G.D., Cisin, I.H. and Manheimer, D.I. (1973) 'National patterns of psychotherapeutic drug use', *Archives of General Psychiatry* vol. 28, pp. 769–83; and Mant, A. (1979) 'Trends in psychotropic drug dispensing', Research Paper 3, in O'Brien (ed.) note 14, pp. 57–65.
28. See, for example, Link, B. and Milcarek, B. (1980) 'Selection factors in the dispensation of therapy: the Matthew effect in the allocation of mental health resources', *Journal of Health and Social Behaviour*, vol. 21, pp. 279–90; and Brown, G.W., Davidson, S., Harris, T., MacLean, U., Pollock, S. and Prudo, R. (1977) 'Psychiatric disorder in London and North Uist', *Social Science and Medicine*, vol. 11, pp. 367–77.
29. Psychotherapy can be just as oppressive as drug therapy. Far from all forms of it are useful. Focus of treatment on psychotherapy alone, no matter what form it takes, is just as reductionistic and oppressive as is drug therapy alone. All levels, which correlate with the 'patient's' behaviour must be changed simultaneously; the cellular, the individual and society itself. Changing the cultural and political conditions of a society to prevent people from being labelled, or rather from being placed in position of being labelled, as 'mentally ill' is an essential level rarely approached for change in Western societies.
30. Dopaminergic transmission is the transmission of electrical activity from one nerve cell to another in nerve pathways which use the chemical dopamine to communicate this information. Acetylcholine is another such chemical which transmits information from one nerve cell to another. This latter transmission is inhibited by anticholinergic drugs.
31. Baldessarini, R.J. (1980) 'Drugs and the treatment of psychiatric disorders', in Gilman *et al.* (eds) pp. 406–14.
32. Crane, G.E. (1973) 'Persistent dyskinesia', *British Journal of Psychiatry*, vol. 122, p. 395–405.
33. Koch-Weser, J. (1977) 'Drug therapy of tardive dyskinesia', *New England Journal of Medicine*, vol. 296, no. 5, pp. 257–60.
34. Terry, R.D. and Davis, P. (1980) 'Dementia of the Alzheimer type', *Ann. Rev. Neurosci.* vol. 3, pp. 77–95.
35. Chapman, S. (1979) 'Advertising and psychotropic drugs: the place of myth in ideological reproduction', *Social Science and Medicine*, vol. 13A, pp. 751–64.

36. Mant, A. and Darroch, D.M. (1975) 'Media images and medical images', *Social Science and Medicine*, vol. 9, pp. 613–18.
37. Smith, M.C. and Griffin, L. (1977) 'Rationality of appeals used in the promotion of psychotropic drugs. A comparison of male and female models', *Social Science and Medicine*, vol. 11, pp. 409–14.
38. See for instance, Lennard, H.L., Epstein, H.J., Bernstein, A., and Ransom, C. (1971) Mystification and Drug Misuse, New York: Jossey-Bass, p. 44; and Hemminiki, E. (1975) 'Review of the literature on the factors affecting drug prescribing', *Social Science and Medicine*, vol. 9, pp. 111–15.
39. Rogers, L. (1982) 'The ideology of medicine', in Rose, S. (ed.) *Against Biological Determinism* London: Allison and Busby.
40. Baldessarini, note 31, p. 439.
41. Klass, A. (1975) *There is Gold in Them Thar Pills*, Harmondsworth: Penguin.
42. Hayes, W. (1976) 'Medicine in conflict with evolution', in Diesendorf, (ed.)note 13, pp. 49–60.
43. Klass, note 41.

5. Western scientific medicine: a philosophical and political prognosis

1. Kuhn, T. (1959) *The Copernican Revolution*, New York: Random House, chapters 3–4. *Cf.* Cohen, I, (1970) *The Birth of a New Physics*, London: Heinemann, Chapter 2. For the concept of a 'paradigm' see: Kuhn, T. (1964) *The Structure of Scientific Revolutions*, University of Chicago Press.
2. Dijksterhuis, E. (1969) *The Mechanization of the World Picture*, Oxford University Press, pp. 386–491; T. Kuhn *op. cit.*, 1959 Chapter 7. For a related discussion of the concepts of law and causality see, Buchdahl, G. (1969) *Metaphysics and the Philosophy of Science*, Oxford: Blackwell, chapter 2. *Cf.* Benton, T. (1977) *Philosophical Foundations of the Three Sociologies*, London: Routledge and Kegan Paul, chapters 3–4.
3. Gay, P. (1973) *The Enlightenment, an Interpretation*. London: Wildwood House, chapter 3–4.
4. Miller, J. (1978) *The Body in Question*, London: Cape, chapter 5. *Cf.* Harvey, W. 'On the motion of the heart and blood in animals', in Clendening L. (ed.) (1960) *Source Book of Medical History*, New York: Dover, pp. 152–70.
5. Descartes, R. 'Meditations on the first philosophy in which the existence of God and the distinction between mind and body are demonstrated', in Haldane E. and Ross G. (eds. and trans.) (1967)

The Philosophical Works of Descartes, Vol. 1, Cambridge University Press, pp. 144–99.

6. For an interesting historical introduction to the concept of 'action' in this sense, see Bernstein, R. (1972) *Praxis and Action*, London: Duckworth, especially Part IV. *Cf.* Taylor, R. (1966) *Action and Purpose*, Englewood Cliffs: Prentice-Hall.

7. Two interesting recent examples are: Dennett, D. (1979) *Brainstorms*, Hassocks: Harvester Press, especially Part IV; Rorty, R. (1980) *Philosophy and the Mirror of Nature*, Part I Princeton University Press.

8. Miller, note 4; also see pp. 216–28. *Cf.* Wightman, W. (1971) *The Emergence of Scientific Medicine*, Edinburgh: Oliver and Boyd, chapters 2–4. Also *Cf.* Cartwright, F. (1977) *A Social History of Medicine*, London: Longman, chapters 1–2.

9. Reiser, S. (1978) *Medicine and the Reign of Technology*, Cambridge University Press, chapter 1.

10. Griggs, B. (1981) *Green Pharmacy*, London: Jill Norman and Hobhouse, chapters 3, 10. Also see excerpts from Molière, *Love's the Best Doctor* and *Le Malade Imaginaire* in Clendening (ed.) note 4, pp. 221–28. For some of the horrors of surgery, pre-anaesthesia and antisepsis, see, Youngson, A. (1979) *The Scientific Revolution in Victorian Medicine*, London: Croom Helm, chapter 1.

11. Reiser, note 9. *Cf.* Dubos, R. (1979) 'Medicine evolving', in Sobel, D. *Ways of Health*, New York: Harcourt Brace, pp. 21–45.

12. Reiser, note 9. chapters 2–5. *Cf.* Youngson, note 10, chapter 4.

13. Reiser, note 9. chapter 6. *Cf.* Dixon, B. (1978) *Beyond the Magic Bullet*, London: Allen and Unwin, chapters 1–3.

14. Jewson, N. (1976) 'The disappearance of the sickman from medical cosmology, 1770–1870', *Sociology*, vol. 10 no. 2, pp. 225–44.

15. For the early development of hospitals, see Foucault, M. (1973) *The Birth of the Clinic*, London: Tavistock; Holloway, S.W.F. (1964) 'Medical education in England, 1830–58: a sociological analysis', *History*, vol. 49, pp. 299–324. For early hospital horrors and a related bibliography, see, Figlio, K. 'The historiography of scientific medicine: an invitation to the human sciences', *Comparative Studies in Society and History*, vol. 19, pp. 274–85. *Cf.* Woodward, J. (1974) *To Do The Sick No Harm*. London: Routledge and Kegan Paul, chapters 9–10.

16. Doyal, L. (1979) *The Political Economy of Health*, London: Pluto Press, chapter 4. *Cf.* Figlio, K. (1978) Chlorosis and chronic disease in nineteenth-century Britain: the social construction of somatic illness in a capitalist society', *Social History*, vol. 3, no. 2; Hodgkin-

son, R. (1973) *Science and Public Health*, Milton Keynes: Open University Press; Stark, E. (1977) 'The epidemic as a social event', *International Journal of Health Services*, vol. 7, no. 4, pp. 781–705.
17. For a general account, see, Smith, F. (1979) *The People's Health*, London: Croom Helm. More specifically, *cf.* Griggs, note 10, chapter 14; Youngson, note 10, chapters 1, 4.
18. Sontag, S. (1978) *Illness As Metaphor*, New York: Farrar, Strauss and Giroux.
19. Doyal, note 16, chapter 2. *Cf.* McKeown, T. (1979) *The Role of Medicine*, Part II, Oxford: Blackwell. Doyal, L. *et al.* (1983) *Cancer in Britain: The Politics and Prevention*, London: Pluto Press.
20. Moss, R. (1982) *The Cancer Syndrome*, New York: McKeown, note 19; Grove Press, chapters 1–5.
21. Macfarlane, A. (1982) 'Saving money, spending lives', *Medicine in Society* vol. 8, no. 1; Blane, D. (1982) 'Inequality and social class', in Patrick, D. and Scambler, G. (eds) *Sociology As Applied to Medicine*, London: Baillierè Tindall, pp. 113–25. For the official view see, DHSS (1980) *Inequalities in Health*, London: HMSO, reprinted as Townsend, P. and Davidson, N. (1982) *Irregularities in Health*, Harmondsworth, Penguin.
22. Dubos, note 11, pp. 26–43.
23. Moss, G. (1973) *Illness, Immunity and Social Interaction*, New York: John Wiley.
24. Eyer, J. (1975) 'Hypertension as a disease of modern society', *International Journal of Health Services* vol. 5, no. 4, pp. 539–58; Eyer, J. and Sterling, P. (1977) 'Stress related mortality and social organization', *The Review of Radical Political Economics*, vol. 9, no. 1, pp. 1–44. *Cf.* Brenner, M. (1979) 'Mortality and the national economy', *The Lancet*, 15 September, pp. 568–73.
25. See for example, House, J. *et al.* (1979) 'Occupational stress and health among factory workers', *Journal of Health and Social Behaviour*, vol. 20, no. 2, pp. 139–60.
26. Illich, I. (1975) *Medical Nemesis*, Part I London: Calder and Boyers.
27. *Ibid.* Part II. *Cf.* Schwartz, G. (1979) 'Biofeedback and the treatment of disregulation disorders' in Sobel, note 11, pp. 359–62.
28. Wintour, P. and Wintour, F. (1982) 'The knives are out', *New Statesman* 15 October. Also see, McKinlay, J. (1979) 'Epidemiological and political determinants of social policies regarding the public health', *Social Science and Medicine*, vol. 13A, pp. 541–58; Salmon, J. (1978) 'Monopoly, capital and the reorganization of the health sector', *Review of Radical Political Economics*, vol. 9, no. 12, pp. 125–33. For a general discussion of cost effectiveness, see,

Cochrane, A. (1972) *Effectiveness and Efficiency: Random Reflections on Health Services*, Oxford University Press.

29. Stocking, B. and Morrison, S. (1978) *The Image and the Reality: A Case Study of the Impact of Medical Technology*, Oxford University Press, for Nuffield Provincial Hospitals Trust. Waitzkin, H. (1979) 'A marxian interpretation of the growth and development of coronary care technology', *American Journal of Public Health*, vol. 69, no. 12.

30. Powles, J. 'On the limitations of modern medicine', *Science, Medicine and Man*, vol. 1, no. 1, pp. 1–30.

31. Doyal, note 16, chapter 6. *Cf.* Ehrenreich, B. and English, D. (1979) *For Her Own Good*, London: Pluto Press; Ruzek, S. (1978) *The Women's Health Movement*, New York: Praeger. For a more general perspective see, Zola, I. (1978) 'Medicine as an instrument of social control', in Ehrenreich J. (ed.), *The Cultural Crisis of Modern Medicine*, New York: Monthly Review Press, pp. 80–100.

32. Three good critical summaries are: Sobel, note 11; Eagle, R. (1978) *Alternative Medicine*, London: Futura; Inglis, B. (1980) *Natural Medicine*, London: Fontana. For a more sympathetic though problematic introduction, see Grossinger, R. (1980) *Planet Medicine*, New York: Anchor Books.

33. For an interesting discussion and bibliography see, Guttmacher, S. (1979) 'Whole in body, mind and spirit: holistic health and the limits of medicine', *The Hastings Centre Report* vol. 9, no. 2, pp. 15–20. *Cf.* Berliner, H. and Salmon, W. (1980) 'The holistic alternative to scientific medicine: history and analysis', *International Journal of Health Services*, vol. 10, no. 1, pp. 133–47.

34. The background to present cuts is analysed in Thunhurst, C. (1982) *It Makes You Sick: The Politics of the NHS*, London: Pluto Press. For an assessment of the positive role of Western medicine against the background of some of the preceding criticisms, see McKeown, note 19, Part III.

35. Porkert, M. (1979) 'Chinese medicine: a traditional healing science' in Sobel, note 11, pp. 147–72; Eagle, note 32, pp. 127–47. *Cf.* (1981) 'How does acupuncture work?' *British Medical Journal* 19 September.

36. Coulter, H. (1979) 'Homoeopathic medicine' in Sobel, note 11, pp. 289–310; Eagle, note 32, pp. 58–78. *Cf.* Shipley, M. *et al.* (1983) 'Controlled trial of homoeopathic treatment of osteoarthritis', *The Lancet*, vol. 1, pp. 97–99; see also, 'The trial of homoeopathy' in the same issue, p. 108.

37. Good introductions to epidemiological methodology are: Alderson, M. (1976) *Introduction to Epidemiology*, London: Macmillan;

Susser, M. (1977) *Causal Thinking in the Health Sciences: Concepts and Strategies of Epidemiology*, Oxford University Press. For a discussion of more general research design, see, Greene, J. 'Experimental design' in Block 3A, DE 304, *Research Methods in Education and the Social Sciences*, Milton Keynes: Open University Press, pp. 47–72. For an excellent introduction to the problem of the 'rationality' of medical science see, Wright P. and Treacher A. (eds.) (1982) *The Problem of Medical Knowledge*, University of Edinburgh Press.

38. They could of course also claim that clinical trials presuppose a reified disease state such as TB or angina whose causes are similar to other cases of the same disease state. Any theory which connects symptoms with the particular peculiarities of an individual person in the holistic sense might not be susceptible to evaluation by this method without profound methodological changes that have not even begun to be formulated – and indeed may not be possible. *Cf.* discussion of humoural theory on p.88.

39. Doyal, L. and Harris, R. (1983) 'The practical foundations of human understanding', *New Left Review*, *Cf.* the discussion of Habermas in Bernstein, R. (1976) *The Restructuring of Social and Political Theory*, Part IV, University of Philadelphia Press.

40. Eagle, note 32, pp. 137–46. *Cf.* (1981) 'Endorphins through the eye of a needle', *The Lancet*, vol. 1, pp. 480–82.

41. G. Schwartz, note 27.

42. Doyal, L. and Gough, I. 'Human needs and social welfare', unpublished paper.

43. Sobel, note 11, Part V. Eckholm, E. (1977) *The Picture of Health: Environmental Sources of Diseases*, New York: Norton. For a broader ecological picture see, Ward, B. and Dubos, R. (1972) *Only One Earth*, Harmondsworth: Penguin. All of these sources should be read against the critical background of, Enzenberger, H. (1976) 'A critique of political ecology', in *Essays in Politics, Crime and Culture*, London: Pluto Press, pp. 253–96.

44. Bull, D. (1982) *A Growing Problem: Pesticides and the Third World Poor*, Oxfam. For many more examples see; Doyal *et al.* note 20.

45. For various examples of how this could work see, *Stress at Work: Final Report*, an investigation carried out by shop stewards of the T & GWU 9112 branch, march 1981; Jones, D., Smith, P. and Kinnersly, P. (forthcoming) *Some Statistical Methods for Assessing Health Hazards at Work: A Guide to Occupational Cancer Studies*, Radical Statistics Pamphlet No. 9; City University Statistical Workshop and GMWU (1983) *Cancer and Work*.

46. Navarro, V. (1980) 'Work, ideology and science: the case of medicine', *Social Science and Medicine*, vol. 14C, pp. 191–205; Renaud, M. 'On the structural constraints to state intervention in health', in Ehrenreich, J. (ed.) note 31, pp. 101–20; Stark, E. (1982) 'What is medicine?' *Radical Science Journal*, vol. 12, pp. 46–89.

6. Human Sociobiology

1.Wilson, E.O. (1975) *Sociobiology: The New Synthesis*, Cambridge, Mass.: Belknap Press.

2. Wilson, E.O. (1978) *On Human Nature*, Harvard University Press.

3. Morris, D. (1967) *The Naked Ape*, and (1969) *The Human Zoo*, both London: Cape; Ardrey, R. (1961) *African Genesis*, (1966) *The Territorial Imperative*, (1970) *The Social Contract*, (1976) *The Hunting Hypothesis*, all London: Collins.

4. Kin selection was first proposed by W.D. Hamilton in 1964 in a now classic paper, 'The genetical theory of social behaviour', *Journal of Theoretical Biology*, vol. 12, pp. 7–32.

5. Wilson, note 2, p. 19.

6. Van den Berghe, P. and Barash, D. 'Inclusive fitness and human family structure', *American Anthropologist*, vol. 79, pp. 809–23.

7. Dawkins, R. (1976) *The Selfish Gene*, Oxford University Press, p. 162.

8. *Ibid.* p. 153.

9. Trivers, R. (1972) Parental investment and sexual selection', in Campbell B. (ed.) *Sexual Selection and the Descent of Man*, Chicago: Aldine, p. 139.

10. Van den Berghe and Barash, note 6.

11. Daly M. and Wilson, M. (1978) *Sex, Evolution and Behaviour*, North Scituate, Mass.: Duxbury Press, p. 275.

12. *Ibid.* p. 279.

13. Barash, D. (1977) *Sociobiology and Behaviour*, Amsterdam: Elsevier, p. 296.

14. Daly and Wilson, note 11, p. 70.

15. In particular see Sahlins, M. (1977) *The Use and Abuse of Biology*, London: Tavistock; Sociobiology Study Group, (1977) 'Sociobiology – a new biological determinism', in the Ann Arbor Science for People Editorial Collective (eds) *Biology as a Social Weapon*, Minneapolis: Burgess; Montagu A. (ed.) (1980) *Sociobiology Examined*. Oxford University Press; Rose H. and Rose, S. (1982) 'Moving right out of welfare – and the way back', *Critical Social Policy*, vol. 2,

pp. 7–18; Dialectics of Biology Group (1981) *Against Biological Determinism*, London: Allison and Busby.

16. Lumsden, C.J. and Wilson, E.O. (1982) *Genes Mind and Culture*, Harvard University Press, pp. 80–81.

17. Lamb, M.E. (1983) Early mother–neonate contact and the mother–child relationship', *Journal of Child Psychology and Psychiatry*, vol. 24, pp. 487–94.

18. Dunn, J. (1979) 'Understanding human development: limitations and possibilities in an ethological approach', in von Cranach, M. *et al.* (eds.) *Human Ethology*, Cambridge University Press; Richards, M.P.M. 'Effects on development of medical interventions and the separation of newborns from their parents', in Shaffer, D. and Dunn, J. (eds.) *The First Year of Life*, London: John Wiley. (1980).

19. Brown, J. and Bakeman, R. (1980) 'Relationships of human mothers and their infants during the first year of life: effects of prematurity', in Bell, R.W. and Sutherland, W.P. (eds.) *Maternal Influences and Early Behaviour*, New York: Spectrum.

20. Van den Berghe and Barash, note 6, p. 815.

21. *Ibid.* p. 811.

22. *Ibid.* p. 814.

23. For example, Herman, J.L. (1981) *Father-Daughter Incest*, Harvard University Press.

24. Barash, D. (1979) *Sociobiology: The Whisperings Within*, New York: Harper and Row, p. 31.

25. Daly and Wilson, note 11, p. 319.

26. Harris, M. (1980) 'Sociobiology and biological reductionism', in Montagu, note 15, pp. 321–22.

27. Sachs, K. (1975) 'Engels revisited: women, the organisation of production, and private property', in Reiter R. (ed.) *Towards an Anthropology of Women*, New York: Monthly Review Press.

28. See for example Harris, M. (1974) *Cows, Pigs, Wars and Witches*, New York: Random House.

29. Mead, M. (1935) *Sex and Temperament in Three Savage Tribes*, New York: Mentor Books.

30. Wilson, note 1, p. 549.

31. Wilson note 2, p. 99.

32. Anderson, P. (1974) *Lineages of the Absolutist State*, London: New Left Books.

33. Lumsden and Wilson, note 16, p. 177.

34. *Ibid.* p. 359.

35. *Ibid.* p. 193.

36. Haraway, D. (1978) 'Animal sociology and a natural economy of

the body politic: a political physiology of dominance', *Signs: Journal of Women in Culture and Society*, vol. 4, no. 1, p.21–36.
37. Leach, E. (1978) 'The proper study of mankind', *New Society*, 12 October.
38. Harpending, H. (1982) *Animal Behaviour*, vol. 30, p. 310.
39. Dawkins, note 7, p 126.
40. Charlesworth, B. (1982) '"Cultural and biological evolution", a review of Cavalli-Sforza and Feldman (1981)', *Quarterly Review of Biology*, vol. 57, pp. 300–03.
41. Barker, M. (1981) *The New Racism*, London: Junction Books.
42. Hirshleifer, J. (1977) 'Economics from a biological point of view', *Journal of Law and Economics*, vol. 20, pp. 1–52.
43. Verrall, R. (1979) 'Sociobiology: the instincts in our genes', *Spearhead*, no. 127.

7. Animal behaviour to human nature: ethological concepts of dominance

1. Schelderaup-Ebbe, T. (1922) 'Contributions to the social psychology of the domestic chicken', *Zeit. Psychol*, vol. 88, pp. 225–62.
2. See, for example, Syme, G.J. (1974) 'Competitive orders as measures of social dominance', *Animal Behaviour*, vol. 22, pp. 931–40.
3. Goodall, J. (1971) *In the Shadow of Man*, London: Collins.
4. Clutton-Brock, T.H., Harvey, P.H. (1976) 'Evolutionary rules and primate societies', in Bateson P.P.G. and Hinde R.A. (eds) *Growing Points in Ethology*, Cambridge University Press.
5. Wilson, E.O. (1975) *Sociobiology: The New Synthesis*, Cambridge, Mass.: Belknap Press, p. 287.
6. See, for example, Rowell T.E. (1974) 'The concept of social dominance', *Behavioural Biology*, vol. 11, pp. 131–34; Gartlan, J.S. (1968) 'Structure and function in primate society', *Folia Primatologica*, vol. 8, pp. 89–120; and Schneirla, T.C. (1946) 'Problems in the bipsychology of organization', *J. Abnormal Social Psychology*, vol. 41, no. 4, pp. 385–97.
7. Bernstein, I.S. (1970) 'Primate status hierarchies' in Rosenblum, L.A. (ed.) *Primate Behaviour*, vol. 2, New York: Academic Press.
8. For example, see Hinde, R.A. (1974) *The Biological Bases of Human Social Behaviour* New York: McGraw-Hill; and Benton, D. Dalrymple-Alford, J.C., Brain, P.F. (1980) 'Comparisons of measures of dominance in the laboratory mouse', *Animal Behaviour*, vol. 28, pp. 1274–79.
9. See, for example, the work of Chase, and that of Landau: Chase,

I.D. (1974) 'Models of hierarchy formation in animal societies', *Behavioural Science*, vol. 19, pp. 374–82; Landau, A.G. (1951a) 'On dominance relations and the structure of animal societies. I Effect of inherent characteristics', *Bull. Math. Biophys*, vol. 13, pp. 1–19; Landau, A.G. (1951b) 'On dominance relations and the structure of animal societies. II Some effects of possible social factors', *Bull. Math. Biophys* vol. 13, pp. 245–62; Landau, A.G. (1965) 'Development of structure in a society with a dominance relation when new members are added successively', *Bull. Math. Biophys*, vol. 27, pp. 151–60.

10. Schneirla, note 6.

11. Crook, J.H. (1970) 'Social organisation and the environment – aspects of contemporary social ethology', *Animal Behaviour*, vol. 18, pp 197–209.

12. Chalmers, N.R. (1981) 'Dominance as part of a relationship', *Behavioural and Brain Sciences*, vol. 4, no. 3, pp. 437–38.

13. Bernstein, I.S. (1981) 'Dominance: the baby and the bathwater', *Behavioural and Brain Sciences*, vol. 4, no. 3, pp. 419–29.

14. For example, Clutton-Brock and Harvey, note 4, p. 206.

15. Rowell, note 6.

16. Bernstein, I.S., Gordon T.P. (1980) 'The social component of dominance relationships in rhesus monkeys (*Macaca mulatta*)', *Animal Behaviour*, vol. 28, pp. 1033–39.

17. Wade, T.D. (1978) 'Status and hierarchy in non-human primate societies', in Bateson P.P.G. and P.H. Klopfer, (eds) *Perspectives in Ethology*, vol. 3, New York and London: Plenum Press.

18. Seyfarth, R. (1981) 'Do monkeys rank each other?' *Behavioural and Brain Sciences*, vol. 4, no. 3, pp. 447–48.

19. Noe, R., De Waal, F.B.M., Hooff, J. (1980) 'Types of dominance in a chimpanzee colony', *Folia Primatol*, vol. 34, pp. 90–110.

20. Crocker, D.R. (1981) 'Anthropomorphism: bad practice, honest prejudice?' *New Scientist*, vol. 91, pp. 159–62.

21. Rowell, note 6.

22. Tiger, L. and Fox, R. (1972) *The Imperial Animal*, London: Secker and Warburg.

23. Bleibtreu, J.N (1968) *The Parable of the Beast*, London: Victor Gollancz.

24. Knipe, H. and Maclay, E. (1972) *The Dominant Man*, London: Souvenir press, pp. 155 and 166.

25. Tiger and Fox, note 22.

26. Mackinnon, J. (1978) *The Ape Within Us*, London: Collins.

27. Knipe and Maclay, note 24.

28. Tiger and Fox, note 22.
29. Anderson, P. (1974) *Lineages of the Absolutist State*, London: New Left Books.
30. Sahlins, M. (1972) *Stone Age Economics*, Chicago: Aldine.

8. Population, poverty and politics

1. One popular best-selling introductory ecological textbook defines ecology as 'concerned with exponential growth, the control of population size, and the structure of communities and ecosystems' and focuses almost entirely on population ecology (Owen, D.F. (1974) *What Is Ecology?* Oxford University Press).
2. See chapter 6.
3. The ideology of population biology presents a complex of related themes encompassing many of the key issues in sociobiology. The discussion here is necessarily limited principally to the narrow aspects of animal and plant population regulation but the whole topic is closely related to the issues dealt with in chapter 9.
4. Malthus, T.R. (1958) *Essay on the Principle of Population*, London: Dent and Sons. The first version of Malthus' *Essay* appeared in 1798 and achieved immediate notoriety. Partly because of its reception, Malthus immediately started work on a second book, travelling extensively abroad in order to collect material. His second *Essay* which was almost four times longer, appeared in 1803 and went through four further editions, in 1806, 1807, 1817, and 1826. There are differences in content between the first and subsequent editions: in the latter, Malthus admitted of the possibility of 'moral restraint' in addition to what he called the 'natural checks' on population of 'misery' (starvation and disease) and 'vice' (wars, infanticide). However these are less important than the differences in presentation. The first *Essay* was written as a polemic against the theorists of the French Revolution. The second was – in appearance at least – a painstaking sociological treatise in which the polemic of the first essay seems to have been deliberately toned down. Nevertheless the arguments are the same. Later, Malthus summarised his ideas in a short essay – *A Summary View of the Principle of Population* published in 1830 – just before the enactment of the New Poor Laws.
5. Malthus, note 4.
6. *Ibid.*
7. *Ibid.*
8. Wallace, A. (1905) *My Life*, Chapman and Hall.
9. This section relies heavily on an excellent paper: Kingsland, S.

(1982) 'The refractory model: the logistic curve and the history of population ecology', *Quart. Rev. Biol.* vol. 57, pp. 29–52.
10. Pearl, R. and Reed L.J. (1920) 'On the rate of growth of the population of the United States since 1790 and its mathematical representation', *Proc. Nat. Acad. Sci. US* vol. 6, pp. 258–88.
11. Hogben, L. (1931) 'Some biological aspects of the population problem', *Biol. Rev.* vol. 6, pp. 163–80.
12. Between 1920 and 1927 Pearl published alone or with Reed more than a dozen articles in a wide array of scientific and popular journals.
13. Interestingly, Lotka's commitment in 1922 to the logistic theory was a consequence of the way in which it fitted in so well with his interest in the 'kinetics of evolving systems' including physical chemistry, just as Robertson on whose work Pearl's logistic was based, derived his theory from a determinist analysis of catalysed reactions.
14. Uvarov, B.P. (1931) 'Insects and climate', *Trans. Entomol. Soc. Lond.* vol. 79, pp. 1–247.
15. Nicholson, A.J. (1931) 'On the balance of animal populations', *J. Animal Ecol.* vol. 2, pp. 132–78.
16. Andrewartha, H.S. and Birch L.C. (1954) *The Distribution and Abundance of Animals*, University of Chicago Press.
17. Lack, D. (1954) *The Natural Regulation of Animal Numbers*, Oxford University Press.
18. Lack, D. (1966) *Population Studies of Birds*, Oxford University Press
19. Ehrlich, P. and Ehrlich, A. (1972) *Population, Resources Environment – Issues in Human Ecology*, Freeman.
20. Ehrlich, P. (1970) *The Population Bomb*, Pan/Ballantine.
21. Interview with David Bellamy in *City Limits* no. 81, 22–28 April 1983. Later, Bellamy declares: 'I have never been political. I have never voted in any election.'
22. Vogt, P. (1948) *Road to Survival*, London: Gollancz.
23. Hardin, G. (1968) 'The tragedy of the commons', in *Science*, vol. 162, pp. 1243–48.
24. *Ibid.*
25. B. Commoner (1972) *Closing Circle*, London: Cape.
26. Paddock, W. and Paddock, P. (1968) *Famine 1975* Weidenfeld and Nicholson.
27. Hardin, note 23.
28. Environment Fund, 'The Real Crisis Behind the Food Crisis' (advertisement) reproduced in *New Internationalist* vol. 79, September 1979.

29. *New Internationalist* vol. 79, September 1979. Susan George's (1976) *How The Other Half Dies*, Penguin, provides one of the best accounts of the realities behind the world 'food crisis'

30. Independent Commission on International Development Issues (1980) *North-South – A Programme for Survival*, Pan Books.

31. Quoted in Mass, B. (1973) 'Rx for the people: preventive genocide in Latin America', *Science for the people* vol. 5, no. 2, March.

32. *Ibid.*

33. Ensenberger, H.M. (1974) 'Critique of political ecology', in Rose, S. and Rose, H. (1976) *The Radicalisation of Science*, Macmillan.

34. Indian blacksmith quoted in Mamdami, M. (1972) *The Myth of Population Control*, Monthly Review Press.

35. Rothman, H. (1974) *Murderous Providence*, Hart-Davies.

36. Commoner, note 25.

37. Frank, A.G. (1970) 'The development of underdevelopment', in Rhodes, R.I *Imperialism and Underdevelopment*, Monthly Review Press.

38. Rodney, W. (1976) *How Europe Underdeveloped Africa*, Bougle L'Overture.

39. Geertz, C. (1963) *Agricultural Involution: The Process of Ecological Change in Indonesia*, University of California Press.

40. *Ibid.*

41. Commoner, note 25.

42. Colvinaux, P.A. (1972) *Introduction to Ecology*, New York: Wiley.

9. Ecology, interspecific competition and the struggle for existence

1. Gilbert, L. E. (1980) 'Food web organization and the conservation of neotropical diversity', in Soulé, M. E. and Wilcox, B. A. (eds.) *Conservation Biology*, Sinauer, chapter 2.

2. Engelberg, J. and Boyarsky, L. L. (1979) 'The noncybernetic nature of ecosystems', *Am. Nat.* vol. 114, pp. 317–24; McNaughton, S. J. and Coughenour, M. B. (1981) 'The cybernetic nature of ecosystems', *Am. Nat.* vol. 117, pp. 985–90; Knight, R. L. and Swaney, D. P. (1981) 'In defense of ecosystems', *Am. Nat.* vol. 117, pp. 991–92.

3. Lovejoy, A. O. (1964) *The Great Chain of Being*, Harvard University Press.

4. Quoted in *ibid.* p. 252.

5. Malthus, T. (1970) *An Essay on the Principle of Population*, Penguin.
6. Paley, W. (1825) 'Natural theology' in *The Complete Works of William Paley*, vol. 3, London.
7. Chambers, R. (1860) *Vestiges of the Natural History of Creation*, 11th ed, London. For a very readable account of Chambers's book and its reception see Millhauser, M. (1959) *Just before Darwin*, Middletown, Conn.
8. Spencer, H. (1852) 'A theory of population deduced from the general law of animal fertility', *Westminster Review*, vol. 57, pp. 468–501; (1907) 'The development hypothesis', reprinted in *Essays*, New York.
9. The biological parts of his philosophy are found in (1898) *Principles of Biology*, vols I and II, revised edition, London; Williams and Norgate.
10. Smith, A. (first published 1776; this edition n.d.) *The Wealth of Nations*, London: Routledge.
11. An invaluable collection of reprinted material from, and about the period is: Appleman, P. (ed.) (1970) *Darwin*, New York: Norton Critical Editions.
12. Darwin, F. (ed.) (1902) *Charles Darwin*, London: John Murray.
13. This was published in 1858 with Darwin's own views in the *Journal of the Linnean Society, Zoology*, vol. 3, pp. 45–62. Both are reprinted in Appleman, note 11.
14. Quoted in chapter 2. Hofstadter, R. (1955) *Social Darwinism in Modern American Thought*, Boston. This chapter is also reprinted in Appleman, note 11.
15. Carnegie, A. (1900) *The Gospel of Wealth and Other Timely Essays*, chapter 2; reprinted in Appleman, note 11.
16. Quoted in Hofstadter, note 14.
17. Huxley, T. H. (1888) 'The struggle for existence in human society', in *Collected Essays* (1894), vol. 9, London: MacMillan.
18. *Harper's New Monthly Magazine*, vol. 70, pp. 578–90, reprinted in Appleman, note 11.
19. Stoddard, J. L. (1894) *Portfolio of Photographs of Famous Scenes, Cities and Paintings*, Chicago: Werner.
20. Darwin, note 12.
21. Tansley, A. G. (1917). *J. Ecol.* vol. 5, pp. 173–79; Sukatschew, W. (1927) *Zeitsch. Ind. Abst. Vererb.* vol. 47, pp. 54–74; Clements, F. E., Weaver, J. E. and Hanson, H. C. (1929) *Plant Competition*, Washington: Carnegie Institution.
22. J. B. C. Jackson has analysed these papers in an article in (1981) *American Zoologist*, vol. 21, pp. 889–901. He shows that

competition was not neglected in ecology during the period 1920–49, as has been asserted by at least one recent ecologist (see note 45).

23. Lotka, A. J. (1925) *Elements of Physical Biology*, Baltimore; Volterra, V. (1931) *Leçons Sur la Theorie Mathematique de la Lutte pour la Vie*. Paris; Gause G. F. (1934) *The Struggle for Existence*, Baltimore: William and Wilkins.
Carl Nägeli actually published a paper in 1874, 'On the displacement of plant forms by their competitors' which was ignored but which anticipated the later mathematical treatment given interspecific competition by Lotka and Volterra. See Harper, J. L. (1974) 'A centenary in population biology', *Nature*, vol. 252, pp. 526–27.

24. Gause, note 23.

25. Williamson, M. (1972) *The Analysis of Biological Populations*, London: Arnold, particularly chapter 10.

26. Diver, C. (1944) 'The ecology of closely allied species', *J. Animal Ecology*, vol. 13, pp. 176–78.

27. Darwin, C. (1859) *The Origin of Species*, 1st edn, chapter 3.

28. Elton, C. (1946) 'Competition and the structure of ecological communities', *J. Animal Ecology*, vol. 15, pp. 54–68.

29. Williams, C. B. (1947) 'The generic relations of species in small ecological communities'. *J. Animal Ecology*, vol. 16, pp. 11–18

30. Simberloff, D. S. (1970). 'Taxonomic diversity of island biotas', *Evolution*, vol. 24, pp. 23–47. The history of species/genus ratios and their interpretation actually goes back before Elton's 1946 paper and is discussed in detail by Jarvinen, O. (1982) 'Species-to-genus ratios in biogeography: A historical note', *J. Biogeography*, vol. 9, pp. 363–70.

31. Hardin, G. (1960) 'The competitive exclusion principle', *Science*, vol. 131, pp. 1292–97.

32. Cole, L. C. (1960) 'Competitive exclusion', *Science*, vol. 132, pp. 348–49.

33. Skellam, J. G. (1951) 'Random dispersal in theoretical populations', *Biometrika*, vol. 38, pp. 196–218.

34. Boucher, D. H., James, S. and Keeler, K. H. (1982) 'The ecology of mutualism', *Ann. Review of Ecology and Systematics*, vol. 13, pp. 315–47. This is an important paper which shows just how much of ecology has been neglected and our understanding impoverished by the recent obsession with competition.

35. Hutchinson, G. E. (1959) 'Homage to Santa Rosalia, or why are there so many kinds of animals?' *Am. Nat.* vol. 93, pp. 145–59.

36. Brown, J. H. (1981) 'Two decades of homage to Santa Rosalia: towards a general theory of diversity', *American Zoologist*, vol. 21,

877–88. This paper discusses the developments in ecological theory following note 35 and why only some of Hutchinson's ideas were taken up.

37. An account of competition theory is given in Roughgarden, J. (1979) *The Theory of Population Genetics and Evolutionary Ecology: An Introduction*, MacMillan. A very readable account is given by R. MacArthur, who was one of the main people responsible for the development of 'niche theory', in (1972) *Geographical Ecology*, Harper and Row.

38. Heck, Jr., K. L. (1976) 'Some critical considerations of the theory of species packing', *Evolutionary Theory*, vol. 1, pp. 247–58.

39. The data supporting Hutchinson's ratio are now regarded as suspect for this reason and because their interpretation is complex. The technicalities of collecting and interpreting these kinds of data are reviewed by Harvey, P. H., Colwell, P. K., Silvertown, J. W. and May, R. M. (1983) 'Null models in ecology', *Annual Review of Ecology and Systematics*.

40. Brown, note 36.

41. One of the strongest critics is Simberloff, D. (1982) 'The status of competition theory in ecology', *Annales Zoologici Fennici*, vol. 19, pp. 241–53.

42. An instructive example is the heated argument that has taken place over the importance of interspecific competition in the structure of bird communities on islands. J. M. Diamond (1975) originally emphasised it (pp. 342–445 in Cody, M. L. and Diamond, J. M. *Ecology and Evolution in Communities*, Harvard University Press) and was challenged by E. F. Connor and D. Simberloff (1979) (in *Ecology*, vol. 60, pp. 1132–40) who claimed that chance was far more important. Diamond and Gilpin (1982) (in *Oecologia*, vol. 52, pp. 64–74) rebutted their criticism but had to allow that competition was not the only important factor (Gilpin, M. E. and Diamond, J. M. (1982) *Oecologia*, vol. 52, pp. 75–84). These papers and others are reviewed by Harvey *et al.* note 39.

43. For example in situations where the Lotka–Volterra competition equations predict competitive exclusion, co-existence can be the outcome if density independent mortality is introduced into the model. See Huston, M. (1979) 'A general hypothesis of species diversity', *Am. Nat.* vol. 113, pp. 81–101.

44. de Bruyn, G. J. (1980) 'Co-existence of competitors: a simulation model', *Netherlands J. Zoology*, vol. 30, pp. 345–68.

45. Diamond, J. M. (1978) 'Niche shifts and the rediscovery of interspecific competition', *American Scientist*, vol. 66, pp. 322–31.

46. For an example and a discussion of reductionism in ecology see

Levins, R. and Lewontin, R. (1980) 'Dialectics and reductionism in ecology', *Synthese*, vol. 43, pp. 47–48. These issues are also discussed by Haila, Y. (1982) 'Hypothetico-deductivism and the competition controversy in ecology', *Ann. Zoo. Fenn.* vol. 19, pp. 255–63.

10. The tomato is red: agriculture and political action

1. Gorz, A. (1981) *The Politics of Ecology*, Cambridge, Mass: South End Press.
2. Rosset, P. and Vandermeer, J. (forthcoming) 'The confrontation between labor and capital in the Mid-west processing industry and the role of the agricultural research and extension service', *Rural Sociology*.
3. Interviews with O. Pearson, 10 November 1981, Ithaca, New York.
4. *Sacremento Bee*, 3 February 1942, p. C–6; 4 February 1942, p. 4; 7 February 1942, p. 15.
5. The history of the bracero programme is mainly from Galarza, E. (1964) *Merchants of Labor*, West Santa Barbara: McNally and Loftin.
6. Downs, P., Rice, R. Vandermeer, J. and Yih, K, (1974) 'Migrant workers, farmers, and the mechanization of agriculture: the tomato industry in Ohio', *Science for the People*, vol. 110, pp. 7–114.
7. Schurle, B. W. and Erven, B. L. 'The return-risk tradeoffs associated with tomato production in Northwestern Ohio', *Res. Bull.* 1111, Ohio Ag. Res. and Dev. Center, Wooster, Ohio; Rosset and Vandermeer, note 2; Vandermeer, J. H. 'Mechanized agriculture and social welfare: the tomato harvester in Ohio', unpublished.
8. Friedland, W. and Barton, A. (1975) *Destalking the Wiley Tomato*, Research Monograph 15, Dept. Appl. Beh. Sci., Univ. Cal. Davis, June.
9. Vandermeer, J. (1981) 'Agricultural research and social conflict', *Science for the People*, vol. 13, pp. 5–8, 25–30; Vandermeer, J. (1982) 'Science and class conflict: the role of agricultural research in the midwestern tomato industry', *Studies in Marxism*, vol. 12, pp. 41–57.
10. Lewontin, R. C. (1982) 'Agricultural research and the penetration of capital', *Science for the People*, Jan/Feb.
11. For a more complete analysis of pesticides in general, see van den Bosch, R. (1978) *The Pesticide Conspiracy*, Doubleday.
12. Dickson, D. (1974) *The Politics of Alternative Technology*, London: Fontana, p. 10.

13. Gorz, note 1.
14. Schejtman, A. (1983) 'Campesinado y desarrollado rural: lineamientos de una estrategia alternativa', *Investigaciones Economica*, vol. 164, pp. 115–52.
15. NWAG's statement of principles is as follows:
The New World Agriculture Group (NWAG), is an international organisation which analyses the problems of contemporary agriculture in order to develop and implement alternatives. We base our work on the premise that the recurrent problems of the human condition, including hunger, poverty, disease and war, result from power differences between classes.

Since solutions to agricultural problems are neither wholly technical nor wholly social, we attempt to integrate both technical and social approaches in our research and its applications. We seek to go beyond reforms that do not address the unequal distribution of wealth and power in the world, and we reject approaches to science that pretend to be politically neutral.

As progressive scientists who recognise the political and ideological nature of science and technology, we strive to consciously direct our work in alliance with the oppressed. As an organisation, we also strive to live up to non-hierarchical, anti-elitist, anti-sexist and anti-racist principles. We are not aligned with any political party.

16. Downs *et al.* note 6; Vandermeer (1981) and (1982) note 9; Schultz, B. (1982) 'FLOC update: the struggle continues', *Science for the People*, vol. 14, p. 5; Rosset and Vandermeer note 2; *FLRP Pesticide Manual* (1982).
17. Cardoso, F. H. and Faletto, E. (1979) *Dependency and Development in Latin America*, University of California Press.
18. Rosset, P. and Vandermeer, J. H. (1983) *The Nicaragua Reader: Documents of a Revolution under Fire*, Grove Press.
19. Collins, J. (1982) *What Difference Could a Revolution Make? Food and Farming in the New Nicaragua*, Inst. Food and Dev. Pol. San Francisco; U.S. Nicaragua People to People Work Group, (1982) 'Sandinista economics. Towards a new Nicaragua', *Dollars and Sense*, no. 76, pp. 6–7, 18; Rosset and Vandermeer, note 18.
20. Bendana, G. G. (1982) *El Pejibaye*, Instituto Nic. Cost. Atl.
21. Rosset and Vandermeer, note 2; Schurle and Erven, note 7.
22. Morrish, R. H. (1934) 'Crop mixture trials in Michigan', *Mich. State Coll. Agr. Exp. Sta. Bull.* vol. 256.
23. Risch, S. J., Andow, D. and Altieri, M. A. (1983) 'Agro-ecosystem diversity and pest control; data, tentative conclusion, and new research directions', *Environ. Entomol.* vol. 12, pp. 625–29; Marley,

C. F. and Hardwick, C. (1981) 'Intercropping stretches season', *Soybean Digest*, April p. 40.
24. Schultz *et al.* (1982/83) 'An experiment in intercropping cucumbers and tomatoes in southern Michigan, USA', *Scientia Horticulturae*, vol. 18, pp. 1–8; Vandermeer, J. *et al.* (1983) 'Over-yielding in a corn-cowpea system in Southern Mexico', *J. Biol. Ag. and Hort.* vol. 1, pp. 83–96; Rao, M. R. and Willey, R. W. (1980) 'Evaluation of yield stability in intercropping studies on sorhgum/pigeonpea', *Experimental Agriculture*, vol. 16, pp. 105–16.
25. Hansen, M. and Risch, S. (1979) 'Food and agriculture in China. Part II', *Science for the People* vol. 11 no. 4 pp. 33–38.
26. Wells, M. (1982) 'Political mediation and agricultural co-operation; strawberry farms in California', *Econ. Dev. and Cul. Change*, vol. 30, pp. 413–32.
27. Yih, K. (1983) 'Intercropping vegetables', *The Grower*, vol. 16, pp. 12–16.
28. *FLRP Pesticide Manual* (1982).
29. Strong, R. D. (undated) *Integrated Pest Management is Cost Effective*, Assoc. Applied Insect Ecologists, 10202 Cowon Hts. Dr., Santa Ana, CA. 92705; Oatman, E.R. and Platner, G. R. (1978) 'Effect of mass releases of *Trichogramma pretiosum* against Lepidopterous pests on processing tomatoes in southern California with notes on host egg population trends', *J. Econ. Entomol.* vol. 71 no. 6 pp. 896–900.
30. Ann Arbor Science for the People Food and Agriculture Group 'New method of pest control creates jobs', *Nuestra Lucha*, 1980, December.
31.Levins, R. (1981) *NWAG Position Paper.*

11. 'They're worse than animals': animals and biological research

1. Cited in Brown, A. (1974) *Who Cares for Animals: 150 Years of the RSPCA*, London: Heinemann.
2. Banks, O. (1981) *Faces of Feminism*, Oxford: Martin Robertson, p.82.
3. Brown, note 1.
4.Ryder, R. D. (1981) 'British legislation and proposals for reform', in Sperlinger, D. (ed.) *Animals in Research*, Chichester: Wiley.
5. Part of the problem is that, for the purposes of Home Office returns, an 'experiment' consists in the procedures that may be carried out on a single animal. Thus, an animal receiving repeated injections may be counted as one experiment. On the other hand,

an animal having a single injection of something innocuous will be counted as an experiment occurring without anaesthesia; hence the argument that *x* per cent of animals did not receive anaesthetics is somewhat misleading since it includes animals for whom pain and stress were not great and for whom anaesthesia was not required. Also see Dawkins, M. S. (1980) *Animal Suffering: The Science of Animal Welfare*, London: Chapman and Hall, for discussion of other points about defining 'experiments'.

6. What is significant, of course, is that only certain species warrant this kind of attention. The use of dogs, for example, arouses considerable anger (e.g. 'smoking beagles'), whereas there is never any hue and cry about rats and mice. I doubt if rats suffer any less than dogs in experiments, however.

7. Harrison, R. (1964) *Animal Machines*, London: Stuart.

8. See Esling, R. W. J. (1981) 'European animal experimentation law', in Sperlinger (ed.) note 4.

9. See for example, Brown, P., and Jordanova, L. (1981) 'Oppressive dichotomies: the nature/culture debate', in *Women and Society: Interdisciplinary Essays*, London: Virago.

10. Lovejoy, A.O. (1936) *The Great Chain of Being: A Study of the History of an Idea*, New York: McGraw-Hill.

11. Merchant, C. (1982) *The Death of Nature: Women, Ecology and the Scientific Revolution* London: Wildwood House.

12. *Ibid.* chapters 5 and 6.

13. Cited in Easlea, B. (1981) *Science and Sexual Oppression*, London: Weidenfeld and Nicolson, p. 71.

14. See Ryder, note 4.

15. There are various organisations which encourage the development of alternatives such as the Fund for the Replacement of Animals in Medical Experiments (FRAME). At present, alternatives are rather few, although the numbers of animals used could be reduced by using some alternatives, such as tissue culture and other *in vitro* methods. See A.N. Rowan (1981) 'Alternatives and laboratory animals', in Sperlinger (ed.) note 4.

16. Diamond, C. (1981) 'Experimenting on animals: a problem in ethics', in Sperlinger (ed.) note 4.

17. Another part of this argument for some proponents of this view is that these are qualities not necessarily possessed by all human beings. As Diamond notes; 'there is *no* feature of human beings, such as rationality of such-and-such a level, which is actually shared by *all* human beings and *no* other animals, and which could be used as a basis for giving human beings a specially privileged position in morality', *ibid* p. 349.

18. It should be noted, however, that some of this extremely painful animal testing is in principle unnecessary. It would be quite possible to design industrial processes so that they are *safe*, and so that the human worker does not come into contact with a potentially toxic compound. But of course to do so would cost money – more money than capitalist industry is ever prepared to pay. It is cheaper in the long run to risk the workers' health and if necessary to pay if sued for damages. The LD_{50} test has been condemned by many as (a) unnecessarily cruel, and (b) not very informative for human welfare. See Rowan, note 15.
19. Merchant, note 11, p. 190.
20 See, for example, Rose, H. (1982) 'Making science feminist', in Whitelegg E. *et al.* (eds) *The Changing Experience of Women*, (1980) Oxford: Martin Robertson; Arditti, R. 'Feminism and Science,' (1980) in Arditti, R., Brennan, P. and Cavrak S. (eds.) *Science and Liberation*, Boston: South End Press.
21. Arditti (1980) *ibid* pp. 366–67.

Conclusions

1. E.g. R. Lewontin (1979) 'Work collectives: utopian and otherwise', *Radical Science Journal*, vol. 8, pp. 133–37.
2. The Dialectics of Biology group met for an initial conference in Bressanone, Italy, in 1980 and a smaller group have now resumed meetings in London, in 1984. The papers from the conference were published in 1982, under the general editorship of Steven Rose, as *Against Biological Determinism* and *Towards A Liberatory Biology*, London: Allison and Busby.
3. Some of these issues formed the basis of a conference on 'The Biology of Co-operation', which was one of a series organised by the Gramsci Institute, Parma, Italy, around the theme of 'War and Peace: Understanding the Reasons for War in the Construction of Peace'. Some of the issues were discussed by Lynda Birke (1984) 'Outgrowing selfish genes', *New Socialist*, no. 16, pp. 40–42.

Further reading

The following is a short bibliography to enable those who are interested to read further on the topics covered by this book. More specific references can be found in the notes to each chapter. Many of the following titles are edited collections of articles, covering a number of topics on the theme of radical critiques of science and its practice. In addition to these are journals, such as *Science for People*, produced by the British Society for Social Responsibility in Science (BSSRS), its US equivalent *Science for the People*, and *Radical Science Journal*. Articles on science also appear in various magazines, such as *Undercurrents, Spare Rib, New Socialist* and many others.

Albury, D. and Schwarz, J. (1982) *Partial Progress: The Politics of Science and Technology*, London: Pluto Press.

Arditti, R., Brennan, P. and Cavrak, S. (1980) *Science and Liberation*, Boston: South End Press.

Barker, M. (1981) *The New Racism: Conservatives and the Ideology of the Tribe*, London: Junction Books.

Brighton Women and Science Group (1980) *Alice Through the Microscope: The Power of Science Over Women's Lives*, London: Virago.

Dialectics of Biology Group (1982) *Against Biological Determinism* and *Towards a Liberatory Biology*, both published in London: Allison Busby.

Doyal, L. with Pennell, I. (1979) *The Political Economy of Health*, London: Pluto Press.

Easlea, B. (1981) *Science and Sexual Oppression*, London: Weidenfeld and Nicholson.

George, S. (1976) *How the Other Half Dies*, Harmondsworth: Penguin.

Hales, M. (1982) *Science or Society? The Politics of the Work of Scientists*, London: Pan Books.

Hubbard, R., Henifin, M.S. and Fried, B. (1982) *Biological Woman – The Convenient Myth*, Boston: Schenkman.

Kelly, A. (1981) *The Missing Half: Girls and Science Education*, Manchester University Press.

Kitcher, P. (1983) *Abusing Science: The Case Against Creationism*, Milton Keynes: Open University Press.

Merchant, C. (1982) *The Death of Nature: Women, Ecology and the Scientific Revolution*, London: Wildwood House.

Roberts, H. (1981) *Women, Health and Reproduction*, London: Routledge and Kegan Paul.

Rose, H. and Rose, S. (1976) *The Political Economy of Science* and *The Radicalisation of Science*, both published in London: Macmillan.

Rose, S., Lewontin, R. and Kamin, L. (1984) *Not In Our Genes*, Harmondsworth: Penguin.

Sayers, J. (1982) *Biological Politics: Feminist and Anti-feminist Perspectives*, London: Tavistock.